GET ON THE AIR...NOW!

A practical, understandable guide to getting the most from Amateur Radio

(*Plus* THE Ham Radio Dictionary)

By Don Keith N4KC

EP

Erin Press

Indian Springs Village, Alabama

Also by Don Keith

The Forever Season
Wizard of the Wind
The Rolling Thunder Stockcar Racing Series (with Kent Wright)
Final Bearing (with George Wallace)
Gallant Lady (with Ken Henry)
In the Course of Duty
The Bear: the Legendary Life of Coach Paul "Bear" Bryant
Final Patrol
The Ice Diaries (with Captain William R. Anderson)
War beneath the Waves
We Be Big (with Rick Burgess and Bill "Bubba" Bussey KJ4JJ)
Undersea Warrior
Firing Point (with George Wallace)
Riding the Shortwaves: Exploring the Magic of Amateur Radio
The Spin
The Road to Kingdom Come
The Ship that Wouldn't Die
THE Ham Radio Dictionary

Writing with Edie Hand:

The Last Christmas Ride
The Soldier's Ride
The Christmas Ride: the Miracle of the Lights

www.donkeith.com   www.n4kc.com

TABLE OF CONTENTS

Introduction: Are you ready to ride the shortwaves?

What this book is

Let me make an assumption here. If you are reading this introduction, you fall into one of the following four categories:

- You are interested in the hobby of Amateur (or "Ham") Radio but have not yet started the process of getting an FCC license and putting a station on the air. You may still harbor the fear that even if you became licensed you would not be able to put together a station and actually start sending our radio-frequency energy. And even if you did, you are not sure what you would do or how you would talk with people.

- You have gone to the work and study to get a license, now have a call sign, but for whatever reason have not yet gotten on the air, or the only operating you have done is with a little FM walkie-talkie ("HT") on the VHF/UHF bands.

- You have been a Ham for a while now, may have even put some kind of station on the air, but decided it was too much trouble or your experience with the hobby was less than you expected it to be. Still, you think you might enjoy it all if it was not quite so complicated to get on the air and start communicating.

- You are an active, experienced Amateur already and love the hobby. However, you know others who fit into one of the three groups above and would like to have something to which you could refer them for more help than you are able to offer. Or simply want a book to hand to them.

If my assumption is correct, then this book is for YOU!

I have been a licensed Amateur Radio enthusiast...emphasis on ENTHUSIAST...for more than a half century. I passed my first exam at the age of 13. Though there have been times when I was inactive on the bands, I have always come back to the hobby, embraced its changing landscape, and enjoyed it just as much as I ever did.

Not only has the hobby given me much pleasure and satisfaction, hours of enjoyment, a chance to help my neighbors through public service, and a constant source of new things to learn and explore, it also led directly to a twenty-two year career in broadcasting and an even longer time in advertising, marketing and writing.

Could I say the same thing if I had chosen stamp collecting, fishing, golfing or model-airplane flying as an avocation? Nothing against those fine hobbies, but I don't think so.

Throughout my years in this fascinating past-time, I have had the opportunity to watch many, many folks decide they wanted to become an Amateur Radio operator. Some simply lost interest when they found out they had to learn very basic electronics and a few rules and regulations. Frankly, if their interest was that lukewarm, the hobby was not for them anyway.

Others got a license but soon fell by the wayside. In some cases, they discovered it just was not for them or not what they expected and moved on to something else. But all too often they just became frustrated with the "what comes next" part of any pursuit: putting a station together, getting some kind of antenna up, learning the jargon, and making that first scary contact.

Avoiding the "HT trap"

Before we talk more about putting your station together and getting on the air, let me get something off my chest. I have seen this phenomenon over and over and it saddens me.

You get excited about Ham Radio, study hard, and get a license and call sign. Immediately you are faced with the challenge of how to get on the air. What do you buy? How can you possibly put up an antenna? What would you say to all those experienced operators out there if you did manage to get a working station on the air?

If you start with the Technician-class license—as most people do—and thus have limited HF operating privileges ("HF" means the shortwave frequencies assigned to Hams as opposed to the Amateur bands on VHF and UHF. And the last half of this book is a very complete Amateur Radio dictionary, so I urge you to look up as you go any terms that are not familiar to you.). Why should you even think about a radio that covers all those frequencies you cannot yet use and worry about an ugly outside antenna? Shouldn't you even avoid purchasing a 50-watt, $300 VHF/UHF FM radio until you see if this hobby is really all it is cracked up to be? You just need a radio that will allow you to talk through the "repeater" stations around town. That would let you get your feet wet and get you on the air...NOW!

Besides, you reason, the catalogs and radio store web sites are packed with those little walkie-talkie things for sale. And some of those are priced at less than fifty bucks! That's less than a family night out at Appleby's and, with one of those little hand-held radios, you would be ON THE AIR, using your license, testing the waters in your new hobby.

So okay, you decide, that's it. Easy decision. You want to get on the air quickly. The little HT comes complete with antenna, five watts of FM power, and a built-in battery and drop-in charger. Nothing else to buy. With one of those, you are on the air with no muss and fuss. No surreptitious antenna-raising when the neighbors are not watching, no major ding to the credit card for a big old multi-band/multi-mode transceiver, no steep learning curve while trying to figure out all those meters, buttons and knobs on a big, expensive radio.

I understand the reasoning. The day your new call sign pops up in the FCC database, you can turn on that little handheld device and start yakking on the local repeaters. Ham Radio is yours and you can start enjoying it without fear.

Those little HTs are great and with the entry into the market of the Chinese manufacturers, they are available at so-what-if-I-lose-it prices. Everyone should have at least one walkie-talkie for local repeater use and to assist in public service events, storm spotting, and emergency situations. They are handy for monitoring repeaters, too, no matter where you might be, especially early on in your Ham career as you learn protocol and who is who.

But they are a trap, I tell you. A trap! And here is why.

You finally see your call sign in the database, get excited, pull out the little $50 radio you have already been listening to, and you summon up the courage to make a call. When you let up on the button you hear the squelch tail of the repeater for a few seconds, and maybe the voice or Morse code identification. The radio works!

But nobody responds. You try several more times over the next few days. No answers. Where are all those friendly voices you have been hearing?

Finally, afraid you have gotten a busted radio, you decide to break into an ongoing conversation you hear, just to see if they can pick you up.

"Joe, sounds like somebody trying to get into the repeater. Try it again, Old Man."

(No insult intended there, by the way. Hams call each other "Old Man" all the time if they don't know the other guy's name. Don't ask why. Just accept it. Unless you are female. Then you can get mad about it.)

So you push the PTT ("push-to-talk") button and give your new call sign, your name, your location, and ask them for a signal report.

"Sorry, friend, you just are not making it into the repeater," comes the response. "Maybe try again later when you are in a better location."

And that happens to you a couple more times. Better location? You are hearing the repeater's signal just fine. And this radio is supposed to put out five watts. Maybe it is broken after all.

There is likely nothing wrong with your radio. That is just the nature of VHF and UHF communications. They are termed "line of sight" frequencies. You literally have to be able to see the repeater antenna—or come very close—for the repeater to hear your signal well enough for it to repeat it back by re-transmitting it.

Plus you are most likely using the stock antenna that came with your radio, what Hams call a "rubber duckie." It is a flexible but stubby aerial, made for convenience and toughness, not efficiency or effectiveness. Very few of your five watts are actually being radiated. Instead, they are being used up as heat in that compromise antenna screwed to the top of your HT.

If you plan to use the HT much and are not in a prime position to reach the repeater, you should actually invest in a quarter-wavelength antenna and attach it to your little handheld.

Or if you intend to use the HT in the car, buy an outside "mag-mount" antenna which can be stuck on the roof or trunk deck, run the coax feedline inside, and attach it to the radio. It will make all the difference in the world.

(You are looking up these terms in the dictionary that is part of this book, right?)

Of course, you can also get one of those more powerful FM transceivers, put a more permanent antenna on the car, or put up an outside antenna at home. Note that for home, you will need a twelve-volt power supply. More on that accessory later.

But the fact is that those negative experiences with the cheap HT have dampened your enthusiasm for Amateur Radio more than somewhat. Heck, that cell phone of yours operates somewhere up there in the UHF portion of the spectrum and it runs far less than five watts of power, yet it works. Most of the time anyway.

If this Ham Radio stuff is this persnickety, you decide, you'll just go back to talking to people on the phone and chatting on Facebook. The HT ends up back in its box. You tell friends you just don't understand why folks are so gung-ho on this radio hobby. Shoot, you can't even talk reliably across town with the stuff.

That, in a nutshell, is the HT trap. And I am convinced it robs our avocation of many people who would enjoy it tremendously if they only got past that initial disappointment. That is why I urge people to go ahead and invest in a real station and get at least a marginal antenna up in addition to the cheap HT. There is so much more to the hobby than a handheld and a local repeater, even if you only hold a Technician license.

To get started, get started

That is the purpose of this book, helping those who really would enjoy the hobby and gain from it as I have to get past that initial frustration or reluctance and move on to explore all Amateur Radio has to offer.

It is up to you to decide if the hobby is your cup of tea. I am convinced it is just about the perfect avocation.

I am also just as firmly convinced that the hobby is even more dynamic now that it was when it was first started—when Hams virtually invented radio as a communications medium—just over one-hundred years ago. And I don't care who you are or where your interests lie, you use radio every day of your life, from cell phone to wi-fi to that latest reality television program you watch.

This, in a nutshell, is why I am such a vigorous evangelist for Amateur Radio, and why I hope to convey that same excitement to you. (If, of course, you really want to get on the air and enjoy all the things that the hobby has to offer. If not, feel free to pass this book along to someone else who might have an interest. It could be the biggest favor you ever do for them.)

I acknowledge that the hobby is not for everyone. Even so, I can't help it if I feel sorry for those who do not share our passion for the radio spectrum and the electromagnetic waves, the magical pixie dust that swirls around all of us out there on the wind. To attempt to convey and share that magic, I wrote a book that has become a best-seller titled *Riding the Shortwaves: Exploring the Magic of Amateur Radio*. Even so, I still encounter so many who simply do not understand what it is that attracts us to and makes us so evangelical about our hobby.

Although we Amateur Radio enthusiasts consider ourselves to be communicators, it is really difficult sometimes for us to convey to other folks exactly what it is that makes it so special to us. I was hooked from the very beginning. I still remember my first "QSO" (on-air conversation) after I got my license, put my station on the air with the help of my dad, and made that initial call, a request for someone to answer me and talk for a bit.

That first call is something we term a "CQ"—a blind cast into a mighty big fishpond—looking for someone…anyone…with whom to chat. And, I must admit, I was half hoping nobody would answer. The prospects of such a thing scared me just a little bit.

But when it actually happened, when I made that first contact, it was nothing short of supernatural!

Now, as I mentioned before, Amateur Radio led to a career in broadcasting for me, but I have no doubt I would still love it just as much, even if I had gone a different direction in pursuit of a living wage.

The hobby has helped many people break into and follow careers in communications, electronics, computers, engineering and more. However, many thousands of other Hams have different interests and professions, many decidedly non-technical. You do not have to be a geek to be a Ham!

The hobby came into official being, set up by Congress, in 1912. You don't need much math to realize it is now over 100 years old.

Therefore, it being so old, it should not be surprising to you that some doubt the value of the hobby today. People now have smart phones, video games, the Internet, YouTube, and Facebook. Kids who once might have been fascinated by talking to someone across the continent on the Ham Radio now do it routinely in chat groups, on Facebook, or via texting. Who wants to go to all that trouble to get a license, learn the jargon, hang wire for an antenna, put a station together, risk zapping the picture on the neighbor's big-screen TV, and create radio-frequency energy when all this new planet-spanning technology is available to every one of us?

First, I should point out that the hobby is flourishing. There are more Amateur Radio licensees in the United States today than ever before. Our ranks are at an all-time high and growing rapidly. The rest of the world has also seen resurgence in interest in the hobby in general and even in some of its more antiquated technology, such as Morse code. (And by the way, it is no longer a requirement to know Morse in order to get a license.)

Hams are on the cutting edge of technology, developing and utilizing exciting digital communications just as they did basic radio a century ago. We bounce signals off ionized clouds in the atmosphere, satellites that we have built and launched and that are now orbiting the Earth, as well as off the moon, the tails of comets, and even the Northern Lights. We have integrated advanced computer technology into the hobby, too. Yet we are just as welcoming to non-technical folks as we are to engineers and scientists who just happen to be Hams.

Still, the questions come.

"Can't you do the same stuff on the Internet?"

"I can talk all over the world on my cell phone. I don't need a radio or a tower or all that wire in the sky."

"Why use that Morse stuff when you can just text somebody and tell them what you want to tell them?"

Yes, you can. You can do all those things.

But the people asking those questions are missing the point entirely. It is exasperating to us Ham-nuts that they do not seem willing to even try to understand what draws people to Amateur Radio. And it is vexing to us that we cannot do a better job of explaining the attraction.

In my frustration, I often fire back in reply to such questions with, "Why do you go fishing? You can buy fish at the grocery store. You're a golfer. Why don't you just walk over and drop the ball into the hole instead of teeing it up and whacking at it a bunch of times?

"Tell you what. Take the 'N4KC Challenge.' Pull out that fancy cell phone of yours and dial a number totally at random, using some area code on the other side of the planet. Then see what kind of conversation you have with whoever answers. *If* anybody answers.

"Where's the magic? The cell phone works about the same every time you try it and you have no control over it once you dial a number. There is no thrill in getting somebody to say, 'Hello.' But the ionosphere—that part of the atmosphere that refracts radio signals—changes constantly. When you send out a transmission on the Amateur Radio shortwave bands from your own station, you never know for sure where it will bounce back to earth or who will answer your call.

"However, the fellow on the other end of that enchanted radio circuit will appreciate the magic of what is happening just as much as you do. You will have common ground. Regardless where you were born, where you live, what kind of government you have, what language you mostly speak, or where you are on the this big blue ball we all inhabit, you and that other person out there in the ether are automatically members of the same tribe."

Usually when I go into that rant, the other person just waves me off or changes the subject and I let it go. But sometimes he (or she) will nod or give me some kind of sign of understanding. Then, when I sense the slightest bit of interest, I see it as an invitation to plunge ahead with my zealous message.

"To add to the magic of it all, you may be talking across the country or all the way around the world. That other person might be an astronaut in space, aboard the International Space Station. Most astronauts in the ISS are Hams and often to talk with Amateurs back on Earth.

"Or he or she may be a rock and roll or country music star, an actor in a TV sit-com, a Nobel Prize winner, a best-selling author, or merely a local guy on the way to the supermarket. But that is not all.

"You may eavesdrop on emergency communications in the wake of a hurricane, listen in on someone relaxing in an RV park near Yosemite, hear a network of guys devoted to restoring old motorcycles, pick up someone operating from a mountain summit on solar power or from the shadow of an isolated lighthouse on an obscure island, or overhear a missionary in a South American jungle talking back to family or supporters in the USA.

"Like fishing, you never know what you will catch when you flip on your radio. Like golf, there is a challenge and a sense of satisfaction as you employ learned skills to accomplish something others think of as being difficult to do.

"You are using a radio station you put together and made to work in order to accomplish that miraculous feat. You may have bought the gear—almost certainly did nowadays—and you merely followed the instructions to hook it up and put it on the air. Or you may have soldered the parts onto a circuit board or designed and put up the antenna that is casting those electromagnetic waves off into space. In order to earn your license, you learned enough rules and regulations and basic electronics to pass the test and to put your station on the air. You figured out enough propagation theory to decide on which shortwave band you could most likely conjure up a conversation at a particular time of day. You know enough by that point to put out a clean signal, urge most of the power out of the transmitter and up the feed line to the antenna, and be confident that your signal will be conducted out into space. You acquired enough knowledge to be able to pull a signal out of the atmospheric noise and carry on a conversation using voice, Morse code, the ones-and-zeroes of digital communications, your own Amateur television station, or some other mode you have learned about and now understand how and why it works.

"See, there is fulfillment in that. There is self-satisfaction. Accomplishment. There is…well…magic in the air anytime you are riding the shortwaves!"

Despite being a man of words—a former broadcaster and a writer with over two dozen published books to my credit—and having been a passionate Amateur Radio enthusiast for over half a century, that is about as good a job as I can manage in describing the allure of our hobby to someone who knows little about it.

Maybe the best thing I could say is this: once you catch the fever, you will know it. You will not be able to easily kick it. And you may find it just as difficult as I do to explain it to someone else.

That goes along with being in the "tribe," the "fraternity" (or "sorority") of Ham Radio.

I know it can all seem daunting. A hobby that demands licensing by a federal agency of whatever country in which you may reside? A pastime that requires you to use electricity, radio waves, and other dangerous stuff? An avocation that necessitates some kind of weird outside antenna to really enjoy it to its fullest? One in which you can actually be fined and go to jail if you deliberately break the laws that govern it? And one that has built up over a hundred years of its own esoteric terminology and has a few old-timers who will castigate those who don't know it all when they make that first on-the-air transmission?

Trust me. It is not all that challenging. Kids as young as six years old have passed the entry-level licensing examination. Most people with no knowledge whatsoever of electronics or radio have successfully put stations together and put up antennas without injuring themselves or getting electrocuted. And you will encounter many others on the airwaves who are just like you, new, excited, and ready to learn. And a multitude of others who have been around the block many times and are happy to help newcomers learn and continue to grow in the hobby.

That is what this book is. It is designed to help you get started with your Amateur Radio experience once you pass the exam and become a licensed Ham, ready to tune to a frequency and complete a contact with another member of the tribe.

This book is to help you get the most from Ham Radio by getting on the air...NOW!

What this book is not

At the same time, I will not attempt a soup-to-nuts course that takes you from interested bystander to experienced Ham. That is beyond the scope of this work. Here are some suggestions on taking the first step, learning the basics and passing the licensing examination:

- For licensing courses, visit the web site of the American Radio Relay League (ARRL), the national organization for the hobby in the USA, at **www.arrl.org**. They publish a wide array of books and licensing guides that take the newcomer from zero knowledge to passing the exam with flying colors.

- For even more study materials, and especially if you want to dive deeper into the technical aspects of the hobby, simply use my old friend Google. Search for "Amateur Radio books" or "Ham Radio study guides." There are many available in print, for e-book readers, and in audio and video form. Two excellent providers at reasonable cost are The Gordon West Radio School (**www.gordonwestradioschool.com**) and Dan Romancik's KB6NU "No-nonsense Study Guides" (**www.kb6nu.com/study-guides**).

- While you are Googling, look for a local Amateur Radio club in your area. Simply search for "Amateur Radio club" and your city, county or region. There is also a comprehensive listing of clubs on the ARRL web site. Most are welcoming to newcomers and many offer licensing classes and conduct examination sessions on a regular basis right there in your town. Many also sponsor "hamfests," get-togethers for others just like you who have been infected with the radio bug. (There is something comforting about being in the midst of a group of folks who share your interests and speak the same language as you do, all part of that "tribal" thing I spoke about before. Plus hamfests typically have forums for newcomers, offer the opportunity to shop for equipment for your station, have license exam sessions, and more.)

- Watch for other Amateur Radio operators in your community and approach them with questions you may have, including details about local clubs, hamfests, or licensing classes. You may find Hams operating at a local public event or festival or providing communications for a charity run or bicycle race. Most states also give Hams the opportunity to order Amateur Radio license plates for their cars so you can spot them that way as well. Most are delighted when a potential Ham walks up and asks them about the hobby just because they spotted the auto license plate.

Once you have started down the road to getting your license, you can put to use the suggestions and ideas offered in this book as you jump with both feet right in there amongst the rest of us Hams. Before you know it you will be exciting the atmosphere with your own radio signals, emanating from your very own radio station.

Of course, if you are already licensed and have simply not put a station together yet, let this be the call for you to get on the air…NOW!

Oh, and maybe once you are participating in the magic you will pass this book along to others who might be teetering on the edge. You have my permission to do so. Maybe just this one little shove will be all it takes to push them over and have them join our growing numbers.

Then they, too, can experience the satisfaction, learning, and just plain fun this fascinating hobby offers millions around the world.

(One more reminder: You will encounter some new terms in this book. If you are not familiar with them or can't decipher them in context, you have the most complete Ham Radio dictionary in existence as a supplement to this book. Use it! Use it and learn.

The dictionary is also available as a stand-alone book, great to keep on the desk next to your radio or in the car for quick reference when you hear a new term. And you can get it in old-fashioned paperback or e-book versions.)

Chapter One: Starting to put your station together

Careful! This is really, really complicated so hang with me for a moment.

To receive radio signals, you will need to use a "receiver."

For you to be able to transmit radio signals, you will need to employ a "transmitter."

In the process of both receiving and transmitting, you have to rely on an "antenna."

See, this Amateur Radio stuff is just as dense as you feared!

Of course, I'm kidding. If you have a receiver and a transmitter and they are both hooked to an antenna you are in business and can begin making contacts, working distant stations ("DX"), winning on-the-air contests, meeting fascinating people, and making the most of that hard-won license.

The most important part of your station will be the antenna, but the radio is sort of essential, too. That is where we will start. In the next few chapters we will discuss the various accessories you will either have to have or may want to consider. Then we will get you past the real elephant in the room: the dreaded antenna.

Not transmitters and receivers, now it is transceivers

There was a time when those who wanted a transmitter and receiver had to build them from scratch. Sometimes they had to manufacture their own parts, too. Then, as more people entered Amateur Radio, they began finding used commercial and military surplus radio gear and modifying it. And manufacturers who made that gear saw a new group of potential customers so they started adjusting their military and commercial units to be attractive to the Ham market.

Some companies even started offering kits for those who still wanted to know how their radios were put together and worked but did not want to start with an empty chassis and a schematic diagram. And before long, entrepreneurs started up companies that did nothing else but manufacture equipment for the Amateur Radio market.

The good news: all of this still exists!

You can buy parts and build your own receiver or transmitter based on your own design. With the popularity of the "maker movement," there are plenty of people who love the smell of solder and the challenge of populating a circuit board with capacitors and resistors and making it work.

Quite a few companies and groups offer relatively inexpensive radio kits, complete with all the parts you need and exhaustive instructions on how to wire them all up. In many cases, the kits cost less than you would pay to buy the individual parts. The Internet is rife with chat groups for those kits so finding info and solving potential problems are easily accomplished. However, instructions are so good, most anyone, regardless electronic or construction knowledge can successfully build one of these.

And there are many manufacturers large and small, domestic and foreign, which market wonderfully-designed, ready-to-plug-in-and-turn-on equipment to sate the thirst of eager Amateur Radio operators worldwide.

Remember, too, that with a hundred years of history under our belts, there is plenty of previously-owned equipment out there. Caveat emptor, of course, and we will talk about that more later in this chapter.

All this is to say that you will have no trouble finding equipment with which to get on the air.

I have another bit of good news for you, too, in case you have not noticed it yet. We rarely speak of a transmitter and a separate receiver anymore. Since the 1970s, most gear made for the Ham market has been in the form of the "transceiver." This means both the transmitter and receiver are contained within the same box, and quite often a remarkably small box. Most use transistors, not the old, fragile, hot vacuum tubes of yesteryear, so the transceiver is not so very large and can fit on a small table in a corner of a room, not take up most of a basement as our great-grandfathers' stations once did. That was the reason, by the way, that they were called "boat anchors."

Oh, and there are really small transceivers, too. These are usually what we call "QRP" rigs (we call transceivers "rigs," by the way). This means they run very low output power, typically five watts or less. They are primarily designed for operating from remote locations using battery power but a whole sub-set of Hams enjoy the challenge of operating with QRP power from their home stations.

Remember all that talk in the introduction about the frustrations of low power and compromise antennas? Some in our tribe actually thrive on that sort of self-torture!

But don't those new, high-tech, modern gizmos cost a bunch of money? Yes and no.

You can spend hundreds of thousands of dollars on your Amateur Radio station. Some do. Or you can get on the air for less than a hundred bucks. Some of those QRP rigs I mentioned come in kit form and are as inexpensive as twenty bucks or so! (I don't necessarily recommend that you start your Ham career with a $20 single-band CW QRP rig, but you get my point. No? Well, I'll make it from another angle.)

Think about the cost of a hobby this way. How much does a fishing boat, motor, trolling motor, fish-finder, and a collection of tackle and lures cost? How about a full set of golf clubs, a country club membership, greens fees, and golf cart? Have you priced remote-control airplanes, a decent digital camera, or the basics of getting started in stamp collecting? In comparison, Ham Radio can be far less expensive than other pursuits.

It just depends on what you want to do, where you want to start, and how much you want to spend. Before getting a second mortgage on the house, though, I would suggest you do some serious research.

Decide what you think will be your primary interests in the hobby and choose your rig accordingly. If you are not sure, you may want to opt for something that is more versatile and can grow with you. That does not necessarily mean paying more or that you will buy something so complicated you will grow frustrated trying to figure the dang thing out.

Just as you would with any other purchase, shop around.

Choosing a transceiver

I suggest you start by visiting the web sites of the more prominent manufacturers. See the various transceivers they offer. Shop!

Don't try to understand all the specifications just yet. They are all pretty good. The market is so competitive and on-line customer review sites and chat groups devoted to specific products are so prevalent that any manufacturer that makes shoddy stuff is soon gone. Even the big boys sometimes come out with a lemon but you can be assured that it does not remain on the market very long.

If there is a piece of gear that has been out there for a while, it has a track record. Visit such sites as eHam.net and their "Reviews" section to see what users are saying about specific items. Of course, as with any other customer-review site, you should be cautious and observant. If a transceiver has been out for five years, has over 500 customer reviews, boasts a score of four-out-of-five stars, then gets a 0/5 from some disenchanted chap, I would take that low-ball report with a grain of salt.

Likewise, subscribe to discussion forums on such sites as Yahoo.com. Just do a search for the transceiver models that interest you and I'd wager such a group exists and has many members. Join them. You can easily un-join once you have chosen something else to purchase.

I offer the same caution concerning user forums as I do for customer-review sites. Some who occupy the discussion forums are determined to find something wrong with even the finest piece of gear. Minor glitches or missing features on a radio do not faze some folks while others get apoplectic about them. Others are so proud of their purchase that they can find no shortcomings at all and take offense if anyone else does. Even so, these are great ways to learn about the particular rigs that interest you.

As mentioned before, you may or may not know what aspects of the hobby will appeal most to you. Likely you have no idea right now. Even if you do, they may well change as you grow in the hobby. That is one reason that if I were you I would avoid buying a $5000, do-everything, bell-and-whistle engorged monstrosity of a transceiver to get started.

What if you fall in love with taking a tiny QRP transceiver, a piece of wire, and a battery to mountain peaks? Difficult to haul that Super Whiz-bang 7000 up the trail! Or what if you decide you love VHF and UHF antenna and propagation experimentation? That $5000 radio might work wonders on 20-meter single-sideband but does not cover the part of the spectrum in which you are most interested.

You may also be attracted to the new software-defined radios (SDRs). These boxes interface with your computer and let the software and hardware of the computer do most of the work. They are capable of amazing specs and can be updated frequently with new features and capabilities by simply installing new software. Prices vary from cheap to breathtakingly high. However, if you have to buy a new computer to have something that works with the SDR, or if you don't know a sound card from a CPU, you likely should steer away from this new technology for the time being.

Also, if you expect your primary interest to be storm-spotting or emergency communications, you may just want an FM transceiver that covers the VHF and UHF bands, puts out enough power to reach local repeaters, and has some utility on simplex (standard frequencies reserved for direct contact between stations, not working through repeater stations).

Here are some other thoughts to keep in mind as you look, consider, and buy.

First, as mentioned before, avoid the "HT trap." If you absolutely want to start your Ham career with a walkie-talkie, add a better antenna. For real utility in storm spotting or other emergency or public service activities, you should at least also have a higher power FM transceiver for car or home shack and a good antenna outside the car or house. You do not need an eleven-element beam on a 50-foot tower, but an outside antenna helps tremendously.

Secondly, there are many opportunities to help in emergency situations on the HF bands, 160 through 6 meters. For that—and I know I am stating the obvious here—you will need a radio that will work on those bands. Or at least some of them. There are actually several commercially available rigs that will work on HF, VHF and UHF, and will also allow you to hear and transmit using single-sideband, AM, Morse code, FM, and the various digital modes.

These all-band/all-mode transceivers are dubbed "shacks in a box" because they make it easy for you to operate most available modes of transmission, cover a huge swatch of the available Amateur Radio bands, and are still relatively simple to use. And surprisingly, they retail in the range from just about $900 to less than $2000.

I especially like these rigs because regardless whatever eventually becomes your favorite part of the hobby, you will have a radio that can do it. Maybe not perfectly. Maybe not the best. But they will do it quite well. You can always upgrade later to something newer or better in the particular realm that you prefer, but the "shack in a box" will still be there as a backup. Or you can sell it to help finance your next purchase. These types of radios are coveted by Hams and they are always looking for used ones in good shape and at a reasonable price.

I make no recommendation here but will mention a few. The Kenwood TS-2000 is such a radio. It was one of the first rigs dubbed "DC to daylight" because it literally—in one of its versions—covers most of the Amateur Radio bands from 1.8 megahertz to 1240 megahertz. It also includes the capability of receiving the AM broadcast band, practically all the shortwave spectrum, and many slices of VHF and UHF, too. If you do nothing but listen to shortwave broadcasts, this radio fills the bill nicely. It even has a sub-receiver so you can monitor local VHF repeaters while you chat with someone on 20 meters or elsewhere. They are typically priced in the $1500 range.

Another example is the Yaesu FT-857D. It, too, offers most operating modes, 100 watts output power, and coverage up to 440 megahertz, all in a container about the size of a box of Kleenex. It is an especially well-suited radio for the car or portable operation. You will often see them on sale new for less than $800.

These two rigs have been around for quite a while. That means their technology may not be quite up to the latest but their design and operation have withstood the test of time.

Yaesu has recently introduced a more modern all-mode/all-band transceiver. Their FT-991 contains new features such as the capability of operating digital voice modes and a front-panel touch-screen interface. It typically retails for around $1800.

Not to be outdone—competition within a niche market is wonderful for consumers—the good folks at Icom now have their IC-7100, with a similar set of multi-mode, wide-frequency capabilities although with a decidedly different look and feel. It is usually priced around $1200 per unit and will give you just about the entire realm of Amateur Radio frequencies and modes.

Note that there are some negatives to these "Ham shack in a box" rigs. Since they cram so much capability into one small box, they do not necessarily do everything as well as radios that concentrate on more limited frequency bands or modes. Even so, they are all at least decent at what they do, and actually much better than radios that were top-of-the-line only a few years ago.

Because of their small size and immense abilities they rely on menus within menus for typical adjustments. There simply is not enough room for all the knobs and buttons that would be needed to do all the things these babies can do. This can be a tad confusing, especially to a new user.

Still, the user manuals for all these boxes are well written. Because of their popularity, there are likely to be locals who own them and can be of help if you need it. At the least, there are very large and active discussion groups for each of these radios on Yahoo and other Internet forums.

Again, if you are truly interested in the hobby, I highly recommend starting out with radios that allow you to sample an array of all that Amateur Radio offers. You will quickly settle on those aspects that appeal most to you. If your particular radio or radios help you to discover what those are, you will be far more likely to get the most out of the hobby.

Where do I purchase that wonderful radio?

There are many helpful merchants who offer for sale Amateur Radio equipment from a large number of manufacturers. (Many are listed in the accompanying Amateur Radio dictionary.) They are more than happy to assist a newcomer in making a decision on what to buy to build a first station. In my experience, established vendors are not likely to push something on you that you don't need or take advantage of your naiveté. Word travels quickly in this niche market. eHam.net has a section in their "Forums" pages devoted to customer reports on companies serving the Amateur market. An unscrupulous or shady merchant that takes advantage of new Hams will soon have a damaged reputation and will likely go away.

It is difficult to go wrong by shopping with any of the established Amateur Radio dealers. Some have multiple locations and may even have a store in your area.

Though certainly not a comprehensive list, some of the ones that immediately come to mind include Ham Radio Outlet, DX Engineering, GigaParts, Amateur Electronic Supply, Main Trading Company, R&L Electronics, and Universal Radio. Google "Amateur Radio dealers" and you can find these and more. They all have web sites on which you can shop, see all the radios, compare them, and then call and talk with a human who is, in most cases, a Ham as well.

There are also several merchants who primarily offer an Internet store front. I suggest you stick to those that have been around for a while and have an established track record. These include Quicksilver Radio (qsradio.com), CheapHam.com, HamCity.com, hamstation.com, and more.

It is the 21st century so there are other places that sell Ham gear, too. One you may have heard of is eBay. You can find plenty of bargains there but be certain to check the sellers' ratings. Many people feel quite comfortable buying things on eBay. If you do, consider it another possible source of that radio you've decided is the one you want.

I should mention that some of the major Amateur Radio manufacturers do not typically offer their products through those dealers. They sell directly to customers. Three major ones that come to mind are Elecraft, FlexRadio, and TenTec. You should make it a point to visit their web sites as you shop.

There are also a number of QRP-rig vendors and we really have not discussed that option. Most of them are smaller companies or non-profit groups, and most offer great gear or kits plus stellar customer service.

However, as previously mentioned, I hesitate to recommend that a newcomer initially gets on the air with a QRP (low-power) radio. Just as with the HT and its "rubber duckie" antenna, I fear the experience would be less than impressive. QRP operation is an exciting side of the hobby and can be a tremendous source of satisfaction. On the other side, it can be way frustrating! For only a bit more money, you can get a more powerful transceiver for your station, then opt for QRP later, when you understand more fully the challenges involved as well as how to help overcome them.

If you still want to look that direction, though, I defer once again to Google. Search for "QRP radios" or "QRP transceivers." Or remember, you can always simply turn the power output on most 100-watt radios down to five watts and QRP all you want to.

How about buying used equipment?

So far we have assumed that you want to purchase new equipment. You may prefer bargain hunting on the used market. That option has its positives and negatives.

It is true that used Amateur Radio equipment shows up all the time on eBay, Craigslist, and similar sites. There are also Ham-oriented sites with personal ads galore for goodies that you could use to get on the air. Those web sites include eHam.net, QRZ.com, QTH.com and others.

Be careful! Just as is the case when you buy a used car or a second-hand washing machine, you run the risk of inheriting somebody else's problems. Do not believe anything a seller says unless you know him or her personally. If possible, talk with an experienced Ham you trust to give you his opinion of the bargain you have been eyeing. Of course, if you are technically adept, you may be able to fix any fixable problems and end up with a great deal.

Watch out for scams! As with any online flea market, crooks sometimes advertise equipment they don't even have and will gladly take your money and disappear.

But with all that being said, most sellers on these sites are on the up-and-up and you can find true treasures at trash prices.

How about those hamfests we mentioned before? Most of them have a flea market or what we sometimes call a "boneyard." In fact, for many such events, the used-gear marketplace is the primary reason for their existence. Many attract a good mix of individuals simply trying to clean out the shack or raise cash for an upgrade as well as those who make the rounds of such events, buying and selling used gear as a sideline business. Either could be a good way to equip your station economically.

All the previous warnings apply here, too, but at least you can look at the equipment and confirm it actually exists, is in one piece, and does not smell of intense conflagration. Sometimes the event organizers have a setup that allows you to plug in the gear and see that it at least lights up and makes noise. Again, if you have an experienced Ham you trust, ask him or her to take a look at your prospective purchase and offer an opinion, both on condition and asking price.

On that note, if you have done your homework prior to the hamfest, you have a general idea of what a piece of used gear should sell for and maybe even common problems other users have found in that radio. A price too high or way too low should send up red flags. Of course, you might come back a few hours later and that way-too-high-price may have been reduced to something far more palatable.

Personally, I would never buy a major item from someone at a hamfest or swap meet if the seller refuses to give me a contact telephone number or email address. Most Hams shoot straight with you. Some don't.

Still, flea markets can allow you to assemble a fine station at a reasonable cost. Besides, they can be a lot of fun, too.

It's a great time to be shopping!

I hope I have given you a few things to think about as you decide on the first major necessity for your station. The one thing I want to emphasize is that it is actually hard to go wrong in making this choice. Most modern equipment functions well. Design and components are far more reliable than previously was the case. Most transceivers will do just about anything the operator wants to do. A thorough study of each radio's features will confirm if that is true in your case.

Of course, you may not have any idea which direction your own exploration of Ham Radio will take you. In that case, I suggest you look closely at the "shack-in-the-box" rigs discussed earlier, whether you investigate them through a dealer or on the used market. These rigs can be a part of your shack for the rest of your lifetime and continue to give you a good experience in the hobby.

Never before in the history of Amateur Radio have we had so many different options in putting together a station. There are more manufacturers, more vendors, and more choices for the newcomer, with many offering truly remarkable gear that does so much more than than radios that were previously manufactured for Hams.

Prices, based on the changing value of the dollar, are actually lower than ever, and there is no doubt buyers get much more performance and dependability for the investment made. The Internet has also given us the capability to learn more about the capabilities and specifications of the various offerings but also made it easier to shop for the best deals.

Hamfests may not be quite as plentiful as they once were, but that is because many of them have been replaced by on-line flea markets that allow you to scan and research from your computer, tablet or smart phone. However, those that have survived have become even bigger and better. Odds are, there are several each year within a few hours' drive from wherever you live.

Look, I am fully aware that all these choices might make your head spin, but the second decade of the 21st century is a remarkably good time to be building that Ham station and getting it on the air.

Now get to it! There is a bunch of us out there on the airwaves waiting to get to know you.

Chapter Two: The boneyard (a short story)

We just spoke at length in the previous chapter about how the hamfest flea market could be a good place to stock that Amateur Radio station you are putting together. This (mostly fictional) story reveals some of the things that can occur at a hamfest. You might not know what the various pieces of equipment in the story are, but I think you will get the picture. It all starts on a chilly Saturday morning in a parking lot not that far from you...

Jerry Lowe leaned back in his folding chair as far as he dared without tipping it over. He stretched his aching back. It was almost 8:30 and most of the early "lookers" had already filed past his couple of folding tables, piled high with various Amateur Radio treasures. The hamfest theoretically opened its doors at 8 but the early-birds seeking bargains had begun showing up in the boneyard well before 7, while Jerry was still pulling stuff from the trunk and rear seat of his car, strategically arranging each item on the card tables.

Some of the early guys would be back with lower offers later, he knew. Some he would likely accept, others he would just grin and nod "No!" He needed the room in his shack and shop, but he was not about to give anything away. Gas was $3.60 a gallon. It was true, too, that he needed cash in case he spotted something else he could not live without. This particular fest had a reputation for having good items in its flea market.

Finally, Jerry stood, yawned (the alarm clock had caused a lot of QRM at 4:30 that morning), stretched, and took advantage of the lull to survey the tailgaters lined up across from and to each side of him. He immediately recognized some of the stuff from previous hamfests and swap meets. Some of it still had the same hand-lettered signs Scotch-taped to their fronts with the same exorbitant prices. Others might have been familiar gear but it was on different tables this particular morning. Seemed like some of it got bought and sold over and over, never getting adopted permanently.

The old Kenwood transceiver with the meter hanging by its leads out of the front panel like a gouged-out eyeball. The dirty stack of ancient, rusty military surplus gear that was certainly worth more as scrap than anything else, but absolutely nowhere near the price the OM had on it. The twisted Mosley beam that appeared to have had a violent entanglement with a tree. The nicotine-enhanced Swan 350 with the bent corner on its case signifying a rough landing at some point in its long, long life. The stacks of old *QST* and *CQ* magazines. Even a cache of *73s*, truly a relic from a bygone era.

But then something caught Jerry's eye. A bit of gray front panel and the distinctive Collins logo. A 75-S3 receiver, resting on a table right across the way from him. Jerry felt a tingle run up his spine. He stepped around his table and crossed the row to get a closer look, keeping a sideways view of his own table in case he got a shopper.

"How much you asking for that old Collins receiver?" he asked the man behind the table.

"$750." The man wore a Collins logo belt buckle. Anyone with a Collins logo belt buckle surely knew how to handle the care and feeding of such a fine piece of gear. Unfortunately, the guy was also fully aware of the radio's true value.

"What's wrong with it?" Jerry asked. He did not mind a fixer-upper.

"Just like brand new. I re-capped it myself and used it on the air up until a couple of weeks ago. Never been around cigarette smoke, either. Works as good as the day Art Collins soldered it all together."

Jerry twisted the knobs and tried to get a glimpse inside. The receiver was clean, all right. Knobs and switches were properly nimble and tight. Meter case and dial clear. No signs of smoke or fire or melted parts.

"Yeah, but how much would you take?"

"$750."

Jerry scratched his chin.

"Why are you selling it, then?"

"I've been wanting another KWM-2 since I got rid of my last one," Collins Belt Buckle answered. "I love them things but people keep offering me too much money for them and I can't turn 'em down. I got a KWM-2-sized hole on my shelf, just waiting."

"OK, I have a couple of items on my table I'd have to sell first, but I'd give you $650 cash for it if I do."

The man looked up and down the way to be sure nobody could hear their haggling.

"Tell you what," the man replied in a half-whisper. "You bring me $650 cash before somebody else jumps on her, she's yours. That's how bad I want that KWM-2. And if you get her home and she doesn't work the way I say she will, we'll un-do the deal. Money back guarantee's hard to beat."

The two men shook hands and Jerry retreated to where he was set up, but he could not help gazing back at the receiver he coveted so much. It seemed to be uttering his call sign, calling his name. Carefully, he moved his immaculate Johnson Viking Ranger II transmitter and Hammarlund HQ-170 to a more prominent position on one of the tables and re-did a sign to say, "Complete vintage station: $800." If he could get anywhere close to that price, that beautiful Collins would be his.

Not two minutes later, a tall, slender man wearing a "Know Code" tee shirt ambled by, stopped, came back, looked, and lovingly touched the big VFO knob on the Ranger.

"First radio I ever had, that Ranger," he said quietly, as if reminiscing about a long-ago prom date. "I still do some AM with my new-fangled transceiver, but man, these babies sure sound good on AM. Always wanted me a Hammarlund, too. How much you take for them?"

"$800," Jerry said, without hesitation. "If you've checked, they're worth a couple hundred more. These are in great shape, too."

The man worshipfully moved the VFO knob so the pointer arced up and down the band, turned the receiver's dial to match the frequency, and finally looked up. Jerry did not miss the hunger in Know Code's eyes. The dude wanted this station for his very own.

"Tell you what," the man finally said. "I got an Icom transceiver on my table up the way. If I sell that for what I need to get for it, I'd give you $700 for the pair."

Jerry watched a flock of crows claim a hickory tree at the far end of the flea market. They cackled and cawed and quibbled at each other as they jockeyed for the best perch.

"Okay, it's a deal," Jerry told Know Code. "But I can't hold them for you or guarantee they'll still be here at this price."

"Understood, but the old Icom is priced to sell and I've had some tire-kickers already."

Jerry watched the tall fellow hurry back up the way to a table behind a big pickup truck with a massive screwdriver antenna poking up from a stake hole in its bed. Jerry yawned and then eased back into his chair, then answered a few question from others who stopped by. A couple of them made silly, low-ball offers on the Ranger and HQ-170. He sold a plug and a cable or two, but for only a few dollars, not nearly enough to claim the Collins.

When he had a chance, he kept an eye on Know Code's setup. He really needed the guy to move that Icom so they could do their own deal. Sure enough, in a few minutes, a fellow in a GigaParts baseball cap stopped at Know Code's table. He spent several minutes turning the knobs on what appeared to be an older hybrid (used both tubes and transistors) Icom transceiver on the other Ham's table. They talked, laughed, and talked some more. Finally, GigaParts Cap pointed down the row of tailgaters, gestured affirmatively, and shook Know Code's hand.

Jerry nodded at the fellow as he passed his position but the man walked with a purpose, back to a big SUV with its rear gate up and its back end full of gear. He quickly moved a big linear amplifier from beneath a stack and to a more prominent location. He scrawled something on a sign before taping it to the amp's front.

Jerry knew at once what was going on. GigaParts Cap had to sell the amp to buy Know Code's transceiver. And that would enable Know Code to buy Jerry's transmitter and receiver. And ultimately, Jerry could go over and take possession of the 75-S3. Collins Belt Buckle would have to locate his own KWM-2.

The newly uncovered amp down at GigaParts Cap's setup immediately caught some interest from a guy in an Elecraft jacket. Jerry watched with interest as the two Hams haggled, laughed, and haggled some more. By the time they shook hands—preceded by Elecraft Jacket pointing back toward the far end of the flea market to where he had his own tables set up—the flock of crows had abandoned the hickory tree. Now they were looking for worms and other tidbits in a big field at the far side of the building, beyond where the indoor hamfest activities were being held. The population of the crow flock had grown considerably and so had the noise of their fussing.

Jerry sold a few five-dollar items and an HT with a bad battery—totally disclosed to its new owner, of course—but his attention stayed on Elecraft Jacket as he walked away from GigaParts Cap's place. He followed him all the way to a truck at the very end of the tailgate section. It was piled high with tower sections, rotors, and various antennas and parts. Aluminum and steel sprouted everywhere. Even from that distance, and even with the distraction from the casual lookers at his tables, Jerry could see that the fellow had put a new price on something in his truck, and that it had already attracted some quick attention.

"I'll give you $200 for the Viking Ranger," someone offered, interrupting Jerry's observations.

"I wouldn't sell you the tubes out of it for that!" Jerry responded with a grin.

"How much, then, for just the transmitter?"

"I already have an offer for the set," Jerry told him.

"Shoot, I got three HQ-170s already. I bring one more in, I'll need a divorce lawyer. But I'd love to have the Ranger. How much?" The man pulled his billfold from his hip pocket and opened it. It was obese with Andrew Jacksons.

Jerry looked at the lovely Collins receiver on the table across the way. It fairly gleamed in the early spring sunlight. And down the way, at the truck full of towers, Elecraft Jacket had shaken hands with a fellow in a University of Georgia sweat shirt, and that guy was already double-timing back to a table stacked high with VHF and UHF repeaters and base station antennas.

Things were getting more complicated but they were still stirring.

"Naw," Jerry told him. He could already hear that fine audio spilling out the speaker from the 75-S3. "Come on back after lunch and if I still have it, we can talk."

"Okay, but if I see something else in the meantime…"

As the man reluctantly ambled away, Jerry glanced again at Georgia Sweatshirt. He was shifting around on his table a particularly nice looking repeater so it could more easily be seen by passersby. In no time at all, a couple of guys wearing identical red shirts embroidered with the local club's logo walked up.

Jerry had heard someone mention that the club had been planning on adding a backup repeater for their .98 machine. With any luck, they had just decided on a purchase and it would keep this complex "circle of life" in motion.

There was a sudden ruckus over in the open field. Someone had apparently tossed a half-eaten hamfest hotdog that way and the crows were fluttering, squawking, fighting for the morsel. A couple of kids had joined in the fray, throwing out bits of popcorn, and that only contributed to the noise and confusion.

Meanwhile, Club Repeater Guys were already shaking hands with Georgia Sweatshirt, pointing back toward the table where the club had stacks of donated gear they were trying to sell to enhance the treasury. Obviously, they needed to move an item or two from the club table to be able to afford the backup repeater system.

Jerry tried to tune out the QRM from the flock of crows and looked toward the club's table. Members putting on the hamfest would be working and couldn't man their own flea market tables so they put gear they wanted to sell on the club table. Volunteers stuck with that thankless duty would not necessarily be as aggressive in moving items as a Ham coveting a Collins receiver would be. This snag might derail the whole round-robin that had been so promising up to this point for all concerned.

However, somebody familiar was now standing over there at the club tables, blocking Jerry's view as whoever it was studied a particular item on display there.

Wait. Wasn't that Collins Belt Buckle, the owner of the 75-S3 Jerry desired so badly? The man shifted his position slightly and Jerry could see that he had been lovingly caressing what appeared to be a Collins KWM-2 transceiver. And he was telling the volunteer behind the table something, even as he pointed to where Jerry sat, watching, stretching, holding his ears to block out the screeching of the black birds in the nearby field.

That's when Jerry realized what had to happen.

Collins Belt Buckle was already marching his way, a determined look on his face.

"You still want the receiver or not?" he asked loudly as he drew near, to be heard over the crows. "I just became a very highly motivated seller."

Jerry looked down the way, toward where Know Code had been trying to sell his Icom transceiver. Sure enough, Know Code was looking his way, as if he had figured out what was happening, too.

Jerry pointed toward Collins Belt Buckle and did an exaggerated questioning shrug of his shoulders. You still want the Ranger and Hammarlund?

Know Code gave him the "Wait!" sign with the palm of his hand. Jerry turned and, remarkably, GigaParts Cap was watching them both from the other direction. He gave Know Code a tentative thumbs-up, then the same "Wait!" signal, and turned to look even farther down the way, toward where Elecraft Jacket stood next to his stack of galvanized gravity-defiance sticks.

It took only a casual thumbs-up by GigaParts Cap to attract Elecraft Jacket's attention there at his own table. He nodded animatedly and started jogging up the way to Georgia Sweatshirt's table full of repeaters. The two men nodded and shook hands again and Georgia Sweatshirt trotted over to the club table.

The two Club Repeater Guys were just leaving the club table, bound for the snack bar inside. Thirty seconds later and they would have been gone. Georgia Sweatshirt flagged them down and pointed to each of the other Hams' setups and then, finally, to the club table, explaining to them what was going on. They all enjoyed a laugh.

Thankfully, the flock of crows had flown on, looking to scavenge and pore over morsels somewhere else. Time temporarily stood still as Jerry quickly considered the situation.

For anything to happen, one of the various cogs in this wheel would have to let go some dollars. That would start the chain reaction. But, once the big deal started, if even one of them balked on a pre-arranged deal, somebody in line—and maybe most of them—would be left holding the bag. Or a piece of gear, at any rate.

Jerry stood and walked over to Collins Belt Buckle, took out his wallet, and peeled off enough bills to complete their portion of the transaction. Thank goodness for that last-minute stop at the ATM that morning. He made sure everybody else involved saw that he was starting the snowball rolling downhill, right then and there.

He hauled the beautiful Collins receiver back over and placed it lovingly beneath a blanket in the back seat of his car. He was thrilled, not just at the price but at what appeared to be a fine piece of American craftsmanship that now belonged to him. He couldn't wait to get it home, get it on the table in the shack, and hook up AC and an antenna.

By the time he had the radio properly placed and covered with a protective blanket, he turned around to find Know Code standing there, money in hand, ready to consummate the deal. Jerry thanked him, put the cash into his newly-emptied wallet, and helped complete the transaction by carrying the heavy Ranger transmitter for the buyer back to his truck. Meanwhile, Know Code hugged the big Hammarlund receiver close to his chest, a grin on his face, as they made their way up to his vehicle.

GigaParts Cap was already there, waiting for them. He helped them put the gear into Know Code's truck and, as Jerry headed back to his own table, those two Hams quickly settled up.

GigaParts Cap winked at Jerry as he double-timed back past him, proudly carrying the transceiver he had just purchased. He was on the way to where Elecraft Jacket was already standing at the SUV, lovingly studying the meters on the front of the big amp. Jerry watched as both of them lifted and carried the after-burner down to where the truck full of tower sections and antennas sat.

With the amp safely deposited in the truck's crew-cab, both men helped Georgia Sweatshirt carry multiple tower sections back and lay them down on the asphalt next to his truck. They carefully avoided toting them too close to the precarious mountain of repeaters and antennas, circumventing a costly avalanche.

Club Repeater Guys joined to help in moving the last of the tower sections, and then they consummated the deal with Georgia Sweatshirt for the repeater and antenna, happily toting the pieces over to the club table. There, Collins Belt Buckle was carefully counting out the bills—some of them the very ones Jerry had given him only moments before for the 75S-3—and excitedly collecting his KWM-2.

The sun was now warm on Jerry's face as he eased back down into the chair, and, after the early wake-up that morning and the drive over to the hamfest, he was on the verge of dozing off. He was starting to dream of the 75-S3 and how great it would accent his shack. He already had a manual and had printed out some mods he wanted to try.

Just then, someone walked up to his table, casting enough of a shadow to pull him back from the rapidly approaching dream.

"How much you asking for the FT-1000?" The man had a big "I DO YAESU" button on his hat.

"Tell you the truth, I don't much want to sell it," Jerry told him honestly. "I just don't need it and brought it down to see what it might bring."

Yaesu Button rubbed his chin and made an enticing offer, more than Jerry would have imagined.

"Well, sir, I'd sell it for that, I guess."

Yaesu Button glanced down the length of the boneyard to where a young, blonde lady sat behind a heap of computers, telephones, and odds and ends of various Ham gear.

"Tell you what," the man said. "I need to sell a couple more particular items, and when I do, I'll come back and grab this baby. If you still have it by then, that is."

Jerry thought for a moment. On the trip up to deliver the Ranger to Know Code, he had noticed a particularly nice looking Collins 32-S3 transmitter on somebody's tailgate. He snuck a glance. Yep, it was still there.

A single crow had just settled into the top limbs of the hickory tree. The bird let out a screech eerily similar to some noises Jerry had heard in some DX pileups. Soon, he knew, the tree's branches would be filled with its brethren.

"Okay, here's the deal. I'll be here 'til about 2 o'clock..."

Chapter Three: Must-have accessories for your radio station

As you have likely guessed, in order to get on the air and enjoy that Amateur Radio license you have to have more than a transceiver and antenna. Let's first take a look at the station accessories I believe you will absolutely need, the "must-haves." Then, in the next chapter, we will consider some of the "nice-to-haves," including those I believe are the top five you will want to consider.

Gotta haves

You will be happy to know that there are very few of these. Plus you may have even purchased them already with your radio. Still, there are two things you almost certainly will need and a third that may or may not be a "gotta have."

A 12-volt power supply. I consider this device to be an important part of any Ham Radio station so I will spend a bit of ink/pixels on the subject. Picking the wrong supply can also be another source of frustration that could send a newcomer in another direction in search of a hobby, too, so pardon me if I elucidate a bit here.

The first thing you will notice on your new YaKenIcElTec radio is that there is no power cable to plug into that 110-volt AC wall socket. Practically all transceivers these days require a 12-volt direct current power source. Actually, most rigs prefer about 13.8 volts, but we still refer to these as 12-volt supplies. (The more advanced—read: "expensive"—rigs do have a built-in 110-volt supply and plug into the wall, so be sure to read the specs for the rig you are considering.) Also, most HF/VHF multi-mode radios crave the ability of the supply to handle 20 amperes or so of current flow on a continuous basis. Best to have a supply that will handle the load. Drawing more current than the thing can handle can cause more problems than you want to deal with this early in your Ham Radio experience.

Your radio will draw far fewer than 20 amps on receive and not even close to that on what are termed "intermittent" modes such as SSB. However, for FM or the digital modes, you certainly need 20-amp capability, and it is a good idea to go ahead and invest in a supply that will do that from the beginning. Most mobile/base VHF/UHF FM transceivers require a little more than 10 amps when transmitting. Since you may want to use the supply to power other equipment in your shack at the same time as your rig, I highly recommend that you have at least 20-amp-continuous capability from your power supply.

Note that you can certainly build a power supply if you want. They are not really all that complicated. However, few newcomers have the knowledge or desire to do that. Also, in today's market, by the time you bought the parts—assuming you don't already have a junk box full of electrical and electronic components—you will have spent almost as much money as a good supply costs new and already wired together.

When you begin shopping for a power supply (typically from the same stores you will shop for your radio), you will notice they generally fall into two categories: "linear" and "switching." These labels refer to how these devices convert alternating current (AC) into direct current (DC) and reduce the voltage down from 110 volts in to 13.8 volts output.

The second thing you will notice is that the prices for the so-called switching supplies are considerably less than for the linear jobs. The short reason for that is that the components for the "switcher" are less expensive. Also, the switching supplies are typically smaller and weigh much less.

So, you ask, why is there even a question about which to buy? If they both produce nice, smooth DC, the switching supply is the clear choice, right?

At one time, the switchers were not able to handle higher current and were not especially reliable. They originally became popular for use in desktop computers and they were not under all that much stress in the typical Dell or HP.

They had another huge drawback that kept them from being a good choice for any use with communication equipment. Because of the way the device rectifies AC into DC, they actually create weak radio-frequency signals, typically called "hash." Weak signals but plenty strong enough to render a radio receiver useless when all that hash sizzles and pops in your speakers and covers up stuff you actually want to hear.

Still another issue with some less expensive switching supplies is fan noise. Again, because of how they work, these kinds of power supplies need a fan to move air across components to keep them cool. Some use multiple speeds, changing automatically depending on the temperature sensed inside the box. Some are just downright loud, regardless the speed at which it is running. Loud enough to be darn distracting.

So am I saying to forget the switching power supply and go for the bigger, heavier, more expensive, old-fashioned linear supply with its chunky, hefty transformer? No. Not at all!

Today, manufacturers have found ways to curb all that spurious noise the switching supplies make, or to curb it and put it outside frequency ranges where we are likely to be listening. Likewise, because there is a big market for 12-volt supplies in many applications where fan noise would be objectionable, companies have made great improvements in how loud those cooling fans are.

Again, if you check reviews for the various supplies on eHam.net and other web sites, you can quickly see those that still might be a hash-generator problem, that have cooling fans that sound like jet engines, that may be less reliable, or that have problems handling their specified voltage and current. Ask other Hams, too. But remember, one man's unacceptable sizzle generator and roaring cooling fan is another man's no-problem but cheaper device. The guy may never listen to the 15-meter Ham band, where his supply just happens to spew out enough hash to pin his S-meter. Or he may use headphones all the time and it simply doesn't matter that the supply's cooling fan sounds like a dump truck full of tin cans. Especially if he saved fifty bucks on the supply!

There are plenty of good switching supplies out there. And plenty of solid-as-a-rock and guaranteed-no-hash linear supplies, too. Check the reviews and get whatever is best for you. Keep in mind that if you do end up with a box that has some radio-frequency hash issues or a noisy fan, you can either keep it for a backup and get something else, sell it—with the proper advisement to the purchaser—or use it elsewhere in the shack for other purposes or when you take a station portable somewhere.

Now, there are two more considerations when picking a power supply: meters or no meters and types of output connections.

Some 12-volt power supplies come with a meter or meters on their faces that tell you, among other things, what voltage they are producing and how much current is being drawn by the radio. Some even allow you to adjust the voltage to compensate for the actual current you are getting from the wall socket, the aging of components in the supply, and other factors.

I personally prefer having the meters as well as the adjustment knob. They only add a small amount to the cost of the supply and are good enhancements to have.

(Just be sure that voltage adjustment knob doesn't get jostled and suddenly send too much or too little juice to your nice radio! Don't ask me how I know this can happen.)

Also, as you shop, you will learn that there is a bewildering array of connectors to which you attach the cable from the supply to the radio. Most have a couple of terminal posts—one for positive (typically red in color) and one for negative/ground (usually black). But the box may also have a "cigar plug" output, designed to take the formerly ubiquitous cigarette lighter plug. I actually find this useful sometimes when I want to take a low-power or VHF radio to a park and have access to AC. My supply has a cigar plug output and I keep a cable with a matching plug for quick connect and disconnect.

The supply may also have connections for so-called Anderson Power Poles, a relatively new system on the market that allows reliable but quick hookups. There may be other choices, too, but these are the most common.

My thoughts? The more the merrier. Flexibility is always a good thing. Just keep in mind how you plan to use the supply. If it is going to sit on or under your desk and never move, the binding posts are just fine. If it will be used with a variety of radios in out-of-home locations, more choices might be better.

Be aware, though, that some of those outputs may not handle as much current as others. These are usually clearly marked near the connector. This is so you can easily hook up to the same supply other equipment that may draw less current.

Cost? You can find small 20-amp switching power supplies for about $60. Supplies with metering, variable voltage control, and multiple output connections run from $100 to $150. Linear supplies that can handle 20 amps start at around $200. Supplies, both linear and switching, that have better than 20-amp capability range from $150 and up, depending on how hefty a device you need.

Remember, if you want to operate multiple radios or a shack full of equipment from one supply, it will need to be heftier. Although you likely will not be transmitting on more than one radio at a time, if you keep adding devices to the supply, the total current draw will creep ever upward. Or, if you have the room, you can get three or four "switchers" for the same price as one big linear supply and be covered, even having an extra to go on excursions with you if you so desire.

Microphone. With very few exceptions, your new transceiver comes from the factory with a perfectly useable microphone. It will likely be a handheld mic with a push-to-talk button on the side, already wired so all you have to do is plug it in or screw it onto the radio's mic connector. You can use the PTT or, if it has it, the radio's internal VOX circuit to control the talk-or-listen status of your transceiver.

If you want a microphone that attaches to a stand on the desk or hangs from a mic boom, or if you want to add a foot switch or hand-held switch to turn it on and off, you will need to go shopping again. If you frequent the forums, you will often see posts in which new users ask for recommendations for a microphone that will best match the radio. As you can imagine, opinions are plentiful. Discussion sometimes grows heated.

The truth is, most mics manufactured for the Ham market are good. Models exist that not only match well with the radios but also can be ordered with the proper cable and connector so you do not even have to break out your soldering iron and try to build your own with those tiny wires most mic cables have inside them.

You might also prefer a combination that includes both headphones and a microphone, commonly referred to as a "headset." The same comments in the previous two paragraphs apply there, too.

Morse code key. I hesitate to include this in the must-have section of this chapter. However, if you intend to do Morse code on the bands, you absolutely must have a key! And not a single radio on the market comes with one, as practically all of them do with a microphone.

Note that most radios do come with a built-in automatic keyer, one that makes the dots and dashes for you depending on which way you push the levers on the key. If CW ("CW" is what most Hams call Morse code) is important to you, be sure to verify that the radio does have the internal keyer.

There is a dizzying array of CW keys available for sale out there. If you doubt that the code is still an important part of Ham Radio, just consider how many individuals and companies still offer keys to the market.

The various types available are beyond the scope of this chapter but I will mention two. If you are just starting out, you likely will prefer a simple hand key, often called a "straight" key. You just use an up-and-down lever to turn a switch on and off, making your own dots and dashes. These are relatively inexpensive and some often show up at flea markets for a few dollars.

The key that is used by most Hams is called a "paddle." It uses the internal electronic keyer in the radio or an external keyer—a lot of those on the market, too—to send Morse. You push the finger piece to the right to make as many dots as you want and to the left to create dashes. You can find perfectly good paddles for $50 or so, or purchase others—some of which border on being works of art—for a thousand bucks or more.

Yes, you read that right. You can spend a grand on a device to sends dots and dashes!

Okay, that's it. With a transceiver and a microphone (and antenna, of course), and maybe a CW key, you are ready to communicate on any of the Amateur Radio bands that your class of license allows.

So what are you waiting for? Oh, yeah. Those "nice-to-have" accessories. And an antenna. Or two or three.

Read on and we will get there.

Chapter Four: "Nice-to-have" accessories

Okay, so you have studied the manuals, answered hundreds of exam-pool questions on-line, and happily passed your Amateur Radio licensing exam. Congratulations! You've finally gotten to the point where you can say your mouthful of a call sign without stumbling. And you are shopping for or already purchased your primary station transceiver, the "must have" accessories in the previous chapter, and maybe even something for transmitting and receiving on VHF/UHF in the car. You might also have a decent antenna or two or are actively planning what you will soon erect. You may even be on the air already, making contacts and flinging RF all over the globe like an old-timer.

Wonderful!   Regardless where you are in your Ham Radio experience, now might be a good time to begin thinking about what else you need in your shack to enhance your enjoyment and the fulfillment offered by our amazing hobby. I have some suggestions for you, based on my own opinion, input from others, and my own experience. Others may have different ideas and I assure you they will not hesitate to express them if you ask.

I believe each of the following is "a nice thing to have" but not absolutely required to have yourself some fun and fully enjoy being a Ham. You absolutely do not have to rush out and purchase any of these. Even so, I will list them roughly in order of how desirable I think each is. And again, it is nothing more than my opinion.

Let us start with what I believe should be your top five accessory purchases as you delve into the hobby.

A good watt meter. And by "good," I mean it should be reasonably accurate (lab grade not required) and ideally have the ability to see both forward and reflected power. A cross-needle display is really nice so you can see forward and reflected power at the same time and get a decent idea of your antenna/s standing-waves and standing-wave ratio (SWR).

Those who have seen my previous articles on various web sites (including my own at **www.N4KC.com**) or who have read my book *Riding the Shortwaves: Exploring the Magic of Amateur Radio* know how I feel about SWR. It is a dismally misunderstood but highly over-rated commodity. Still, it is good to have an instant visual check on the integrity of your antenna system. If the SWR on that meter suddenly jumps to 10-to-1 on an antenna that previously presented a 1.2-to-1 SWR, you know something is haywire with your antenna matching device, feed line, or antenna.

And as long as we are talking ideal, I would also highly recommend that you get a watt meter that has a peak-reading function. You may remember from your licensing-exam studies that many watt meters are not able to react in such a way as to allow you to see what actual peak power you are emitting for such modes as SSB.

If you glance at that swinging meter movement over there on a typical average-reading meter—one that does not have a peak-reading function—it may appear that you are only putting out 60 or 70 watts even though you are sure you set the transceiver's output power at 100 big ones. A peak-reading meter will show you more accurately what you are actually doing.

Nowadays, there are several manufacturers that make good peak-reading watt meters with either dual displays or a single cross-needle meter face. Be sure it will handle the power you intend to shoot through it and that it will work on the frequency range for which you will need it. I have seen—and purchased—several such meters, good for HF frequencies, for less than $100 new. Add capability up to UHF and they get pricier, or you may want to just get a separate VHF/UHF meter. Be wary of used meters at flea markets as they may have been scorched beyond repair, but you can also get a good bargain there, too, if you are careful.

I will anticipate another question: Hey, why should I invest in an outboard watt meter—peak or average or whatever—when my spanking new radio has one right there on its face?

Well, that is why this suggestion is in the "nice-to-have" chapter. But I feel strongly enough about the value of such a meter that I make it the first item I suggest here. Believe me, a decent meter will give you far better information about your output power, enable you to better tune your transmitter to your antenna system, and give you peace of mind that all is well with the amount of RF you are shooting toward the heavens.

A dummy load. Yes, a dummy load. Stop airwave pollution while you fiddle with that new radio or tuner. Avoid entertaining the rest of us and the shortwave-listening world while you adjust your transmit audio processing for hours on end with that oft-heard yodel, "H – e – l – l – l - o – o - o – o, radio – o – o – o!"

Not sure what a dummy load is? It is a device that ultimately does exactly the opposite of what your antenna is supposed to do. It provides a good match to the output or your transmitter but then takes the radio-frequency energy it is sending out and squashes it so it is not actually radiated into the sky. They accomplish this task in various ways, but usually eat up all that power by dissipating it in the form of heat, which is precisely what it is supposed to do.

When would you want to do such a foolish thing?

Well, say you, as noted above, want to adjust your transmit audio while listening to your radio's internal monitor function. You could find a clear place somewhere on a Ham frequency and start chattering and yodeling as you tweak mic gain and compression and equalization. But all the while, you are actually sending a signal that may be interfering with someone else. With a dummy load, you can test all you want and not bother anybody.

Again, be sure the dummy load you buy can handle the amount of power you intend to send coursing through its oily innards. Read the manual to be sure you know what the tolerable on-and-off cycle should be so you don't fry its resistors or send its oil bubbling over like a witch's cauldron. And make certain it is rated for the frequencies on which you plan to use it.

A volt/ohm meter. Look, I know not everyone gets into Ham Radio to learn all there is to know about electronics. There is nothing wrong with that, and many other aspects of the hobby besides the technical part attract folks to its ranks. However, a decent volt/ohm meter can serve a number of purposes in your shack, and it is also a nice device for a number of potential household uses, too.

I confess I usually go for the cheapest model Radio Shack or Harbor Freight has on its shelves (using those nice gift cards my kids thoughtfully give me for Father's Day) or something I pick up for a couple of bucks at hamfests. As with the watt meter, you are not looking for something NASA might use.

If you do yearn to learn more and construction and kit-building are in your wheelhouse, invest in something heftier and more fully featured. However, ninety percent of my VOM usage is checking continuity on coaxial cables and jumpers or making sure my 13.8 volt power supply—one of the ones I have that does not have a voltmeter on its plain-Jane face—is somewhere in the general vicinity of 13.8 volts. That and checking the veracity of refugees from that pile of discarded AA batteries. For this reason, I usually prefer an analog meter as opposed to a digital readout.

Take your pick. Or have one of each.

Oh, and remember to check the internal battery in the meter regularly, too. It is frustrating to need to check a coax jumper and find the VOM dead.

An aside here: I found another handy use for those cheap meters. All three of my kids made use of one anytime they had a school science project to complete. Each one did that classic of all science fair demonstrations, "The Conductivity of Various Materials." With the cheap meter tie-wrapped to the cardboard folding display and in the circuit with a 9-volt battery, judges could see which material—including cloth, wood, aluminum, and copper—conducted DC current.

It was a sure grade of B every time!

An antenna switch. Even if you are new to the hobby and only have one antenna out back or stretched through the attic, trust me, you will soon have more. I suspect my aerials are mating back there and having babies. I initially put up a G5RV and a 2-meter/70-centimeter j-pole. Before I knew it, wires crossed my backyard with such regularity that birds could no longer safely fly through the maze.

My rig has two antenna outputs. I have five HF antennas. Simple solution: a 4-position coax switch (my big horizontal loop goes to a second output on my auto-tuner). Read the reviews on eHam.com and other sites on the various switches available. Most are perfectly okay for Amateur use.

You will often find them at flea markets as well. The usual cautions apply. The most common problems are burned or corroded contacts inside the switch, and these are usually difficult to fix. Or not. It is hard to tell. Incidentally, this might be a good use for that cheap volt-ohm meter we just talked about. At least you could quickly tell if the connection on one or more positions on the switch is intermittent or non-existent.

You will see some switches offer a position you can choose that is connected to a ground lug, and that lug can be tied to your station ground. Other switches connect all positions to ground—if one is attached to the lug—if they are not chosen at the time. This is a plus, and can help bleed off a potentially damaging electrical charge on the antennas. This is not, however, even close to an adequate protection against a lightning strike.

Expect to pay up to $100 for a solid, manually-switched device and well above that if you opt for something you remote outside and switch electronically. Either way, you prevent carpal tunnel from constantly screwing connectors on and off. Using such a switch also neatens up the shack considerably.

A computer. I almost left this one off since it is so obvious anymore. But I still talk with Hams who have the computer upstairs and the rig downstairs. When I became active again in 2005, I had my shack in my son's old room—we had finally kicked him out after college—and my writing office/computer in the old family room next door, both in the basement part of the house. I somehow did not realize how firmly the 'puter had become entrenched in the day-to-day operation of a Ham shack.

It did not take me long to move the shack into the office with the transceiver right next to the desktop computer. And I also keep my laptop on the desk, too, so I can interface easily with the radio for firmware updates, memory programming, or even for complete rig control from the computer. I use it to program my HTs and mobile HF/VHF/UHF radio, too.

The main computer is hooked up to the HF rig through a SignaLink USB external sound card for digital modes. And I usually keep DX Summit and QRZ.com open in the browser for quick reference. I love looking up the guy with whom I am ragchewing to see what he has on his QRZ.com page, enabling me to launch into conversation about some of the relevant and interesting things I might find there.

Of course, other logical uses of the computer in the shack include rig control and logging. I came from an era in which we had to keep a paper log. It was required by law!

You cannot imagine the hassle when a QSL card (a postcard-like confirmation) floated in from the bureau and I had to try to find the logbook in which that particular QSO was logged. Or when I figured I had worked up to another DXCC plateau (DX Century Club, an ARRL award for working at least a hundred countries) and tried to track down the various countries to include by riffling through multiple spiral-bound logbooks. Now there are many software packages available—some free, all the rest very inexpensive—that make logging a pleasure.

I am still a casual user of computer rig control. Old-timer that I am, I still kind of enjoy twisting knobs. But it is there when I want to use it, a free software program from my rig's manufacturer. And my radio, as do many of them nowadays, only need a USB cable and the software to communicate with the computer.

There has also been a surge recently in Hams who operate their stations remotely over the Internet. This enables many who live in antenna-restricted neighborhoods to maintain stations at a work location, a lake house, a relative's home, or other location and to fully enjoy the hobby.

You might also want to investigate IRLP and Echolink, two other ways to use your computer as part of the hobby.

Note that a whiz-bang, game-worthy computer is not necessary for most Amateur Radio use. An Internet connection—and a reasonably fast one—is almost a necessity, though.

I suppose if you wanted to pick nits, you could say logging, digital mode, and rig-control software, and the digital interface between computer and rig are additional "accessories," and I have actually named nine here instead of the promised five. Maybe so, but I think you get the point.

That being said, let us now look at some other items you may want to consider for your station set-up, even if they don't fall so neatly into the "gotta haves" and "nice to haves."

This list so far could also have included rather important items like **lightning protection**. Some might maintain the available equipment for that purpose belongs back in the chapter about things you must have.

The possibilities on parts and systems dealing with lightning are beyond the aim of this book. It has been covered in numerous books that deal with that subject alone. I highly suggest you read the sections in the ARRL's *Antenna Book* about how to protect your station from lightning. You can also Google the subject, but be prepared to read many differing opinions.

I do not want to minimize the importance of this topic, nor am I being flippant, but my primary approach to lightning protection is to disconnect my antennas outside the house any time lightning is a possibility.

If you have reasonable surge protection as part of your regular household wiring, that is a good thing. Electrical codes vary so you may want to check with an electrician to be sure you are as protected as you can be.

This brings up another useful accessory for the Ham shack that might even qualify for the "must have" list: a **surge protector**. You do need to be certain the device you employ will handle the current/wattage that the equipment you plug into it will demand.

As with antennas, the only sure-fire way to protect your equipment from nearby lightning strikes or transient power line surges is to disconnect it completely when storms are in the area. That is not just unplugging everything from the AC outlets. It also means disconnecting all the other wires and cables that tie everything together and bring antenna feedlines into the radio. In my case, that is simply not practical. I have over a dozen cables attached to my transceiver and it would be a true hassle to have to disconnect and re-connect all of them every time it thunders.

For that reason, in addition to a good home-owners insurance policy, I also cover my Ham gear, computers, tower, and antennas under the ARRL members' insurance plan. I find it to be reasonably priced and the one time I had to use it, they were a pleasure to work with.

Maybe it is a stretch to say that **insurance** qualifies as an accessory. Be your own judge, but I would not be without it! Think your current homeowner's insurance covers your station and all that goes with it? I suggest you read the fine print on the policy and also check your co-payment amount. Seems less than $50 a year for the ARRL insurance plan—the price depends on how much value you insure—is a small price to pay for that wonderful peace of mind.

Now some other gear that definitely can be called accessories, including some you may truly want, but each can be added in good time and as you grow into the hobby.

A linear amplifier. Yes, they are wonderful to have when the going gets rough, but a good transceiver (which, when I say "good," means a decent receiver with some modicum of filtering and noise suppression) and an efficient antenna system will open up the world for you with 100 watts. Save your pennies and get you an amp someday. I operated without one for the first 47 years of my Ham Radio tenure and have had a blast. Now, I'm often glad I have the extra 9db of signal, but I still would not rate the amp as an essential accessory.

It is absolutely a "nice to have," though.

A tall tower and multi-element HF beam. Again, nice to have, but not essential. And not even an option for many.

A well-designed vertical or wire antenna system will still allow you to work the world. I did it with a G5RV and a multi-band trap vertical. And I've done it with tri-band beams and my current hexbeam. Beams are better than wires but they are not essential.

That being said, many Hams maintain that a dollar spent on an antenna is worth five dollars spent on any other signal-enhancing accessory you might buy. I agree. But I do not see it as an either/or.

It is absolutely true that antennas do not have to cost that much, either. Simple dipoles, G5RVs, or horizontal wire loops are very inexpensive. Even if you decide to buy something commercially available instead of making your own, the prices are typically a hundred dollars or less.

However, if you start pricing towers, the cost of installation, and the money you have to pay for the typical multi-band beam-type antennas, you will realize that you can quickly turn loose considerable capital. All that assumes you can get such a thing approved by the rest of the family unit as well as the homeowners' association and local zoning board.

An **"antenna tuner."** I could make a good argument for this being a required accessory. If you have read previous articles by N4KC and others on antennas, feedlines, and matching devices, then you know why I am an advocate of having at least one of your antennas being a long piece of wire (dipole or loop) fed with open wire balanced feedline so you can use it effectively on multiple bands. If you have gone that route or are thinking about doing so then yes, you must have a "tuner."

The internal auto-tuner available in most radios these days may or may not be robust enough to cover all the bands and their segments on which you wish to dance. However, for the purpose of this chapter, and because not every Ham will absolutely require one, we will keep the antenna matching device in the "nice to have" category. But keep in mind it can also be a must-have gadget, depending on your antenna situation, now and in the future.

Models, features and prices vary greatly. You can buy a reasonably functional matching device ("antenna tuner") that can handle 100 watts and typical antennas and requires you to twist knobs for less than $200. Auto-tuners, boxes that do all the matching by themselves once you feed it RF energy or punch a button, are available for around $300 and up. Auto-tuners that can handle full legal input start in the $500-to-$600 range.

Most of the commercially available matching devices—auto- or manually-adjusted—only work on the HF portion of the spectrum, from 160 through 10 meters.

Incidentally, if you wonder why I add the quotes when I use the words "antenna tuner," it is because this is a truly unfortunate name for these particular devices. That is what most everyone calls them yet they absolutely do not tune your antenna.

What they do is find a combination of capacitive and inductive reactance that shows your transmitter's output circuit an impedance of close to fifty ohms. That allows for the maximum transfer of energy from your radio to the antenna system, which is a good thing.

Sometimes, if the impedance of your antenna system is too far away from fifty ohms, even a "tuner" will be unable to find a match. However, if you are careful to put up an antenna that is reasonably close in its impedance balance between capacitive and inductive, most modern matching devices will allow you to be able to use that antenna on the bands for which it was intended.

Note, too, that the type of feedline you employ to run from your radio/"antenna tuner" to what most people consider to be the "antenna"—the piece of wire or beam in the sky—determines how well you do with a mismatched antenna, brought into workability by a matching device. Coax cable is wonderfully carefree and relatively easy to run into your shack but an antenna fed with the stuff does not perform well if the system has a very high SWR, even if your "tuner" allows you to use the antenna. On the other hand, open-wire, ladder line, or other types of balanced feedlines are a bit of a hassle to use since they must be kept away from other conductive materials. However, even with a high SWR on the line, there is minimal loss and can work well if your matching device can find a match.

This can be a confusing subject to newcomers, and even a controversial one among old-timers. We will talk a bit more about it in the next chapter but for the most part, this book will keep it to a simple maxim: use coax cable if you only plan on using the antenna on the band or bands where you will expect a reasonably resonant match, in which case your radio's internal "automatic antenna tuner" will probably be more than enough. If you want to use an antenna on multiple bands and frequencies, a balanced feedline (open wire, ladder line, window line) will work better but you will likely need a matching device—an accessory, wider-range, outboard "antenna tuner."

You will hear some Amateurs decry the use of tuners. Allow them to vent and then change the subject. "Tuners" are wonderful devices if you know and understand what they do. Depending on what kind of antenna system you have tied to the matching device's output, they can be a tremendous aid in assuring most of the power your radio is producing gets radiated toward the ionosphere and to whomever you are trying to talk with.

And that is not only a wonderful thing, but downright magical.

Chapter Five: Fave five...the top five starter antennas for new Hams

Well, we have postponed this subject as long as we can. As you know by now if you are reading these chapters in order, I firmly believe that the primary stumbling block for newly-licensed Amateur Radio operators is the antenna. They excitedly begin studying for the license, pass the test, begin planning a station, and BAM! They hit the wall when it comes to the antenna.

I understand. I really do. I have been there. Several times.

As a kid, I at least had the help of my dad, a radio/TV repairman who understood the basics of the antenna. I was also blessed with over a hundred acres on which to string some of them.

Then after college, when we moved into our first house, I was pretty much on my own. I remembered enough and did enough studying in those pre-Internet days to manage to hang up some fairly efficient dipoles. But I did have the expertise of one of the engineers at the broadcast facility where I worked when I picked up a used tower and a tri-band trap Yagi beam.

And later, when I became active again after a dozen years making money, spending more money, and raising kids, I found myself fascinated by antennas and was finally ready to do some serious study and experimentation. That was when I started developing my opinions about how others might be able to get over the hesitation, frustration, and disappointment when it came to putting up some kind of antenna and getting on the air.

I realized there were some relatively easy antennas most anyone could put together and get up into the air. I settled on five of my favorites, though there certainly are other options. When it comes to simple and easy, though, these are definitely my fave five antennas that will allow you to get on the air quickly and have a comparatively effective signal for a more satisfying Amateur Radio experience.

The truth is that many new Hams give up too easily. Or are too timid to give something new a try. I have seen several examples of this lately.

One is a new Amateur who got all excited after he upgraded to General class, acquired a perfectly adequate HF station, but so far has not gotten around to erecting any kind of decent antenna to use with all those new privileges and Ham gear.

I don't know if it is because he is intimidated due to a lack of knowledge about antennas or if he simply is not sure what type of antenna to put up. Maybe he hesitated, thinking he should wait until he had the perfect choice, or something that would elicit "ooohs" and "aaahs" from stations he worked. I suspect, though, that he only suffers from the "I don't know what I don't know" syndrome. All this stuff becomes a lot to take in at the beginning.

It can be daunting! But it does not have to be.

I also know of a long-time Ham who came back from a period of inactivity, dragged the old gear out of the closet, and then, for whatever reason, never quite got around to the most important part of the station—the antenna! He threw some wire out the window but could hardly hear anything, and his radio just hissed at him when he tried to tune up to that mess.

I understand his thought process perfectly. I confess that I, too, am a procrastinator. I tend to spend a long time getting ready to start to begin to commence to think about launching a project until I inevitably forget what it was I wanted to do. And, by the way, what did I buy those parts and rope and wire and fiberglass and aluminum for in the first place? Maybe I'll just try something quick and dirty to get on the air and think and plan some more so that whatever I eventually decide on and put up between two trees is perfect.

What we have here are otherwise intelligent people who are enthusiastic about starting or resuming the hobby. However they are allowing fear or hesitancy to keep them on the sidelines. I'm afraid that some of us who attempt to help only contribute to the problem by pushing antenna ideas that are beyond the means, knowledge or geography of the hesitant Ham. Or even sometimes beyond their interest level or desire for learning.

Let's get one thing straight. Not everyone wants to be an antenna engineer. They just want to work some DX or conjure up a ragchew. There is nothing wrong with that!

On the other hand, if you decide you enjoy playing with antennas and propagation of radio signals, man have you found the right hobby! But there is plenty of time for that later. For right now, let us talk about getting you past any trepidation you have about antennas and on the air...now!

In that spirit, here are my five best get-on-the-air-quickly-and-easily antenna ideas. Maybe you have other suggestions, but understand that I am applying the following logic in picking these particular ones:

- They are easy to build for most anyone who is willing to try and do not require any special tools or test equipment.

- They may be crafted from easily available materials and cost very little, so there is not much downside if you mess them up.

- They are not necessarily the be-all, end-all of RF radiators but they do work well enough to give a good experience to the user.

- They are not necessarily the best for all situations, including for use in antenna-restricted neighborhoods or in condos and apartments. That's another book. And the ARRL publishes one on the subject. Check it out.

- And if someone attempts to construct one of these bad boys, he or she will possibly learn a little antenna theory by osmosis and, just maybe, will become curious enough about the subject to learn more and become inspired to try more challenging projects.

Now, in no particular order of preference, here are N4KC's Top Five Get-on-the-Air-Quickly Antennas:

#1 – The half-wavelength-long wire dipole. This one, boys and girls, it is about as basic as it gets and it can work quite well on any HF band. It consists of two pieces of conductive wire—usually some kind of relatively strong copper—and they are stretched end-to-end, joined together in the middle with a short insulator between them. We call that middle area the feed point. Insulators and lengths of rope are attached to each of the opposite ends to support the antenna. Some also call this a "flat top" antenna or a "doublet."

It can be hung between two supports—often trees—parallel to the ground. It can also be supported in the middle with the ends sloping downward in an inverted "vee" configuration. For some reason, this way of doing it is often called an "inverted vee" antenna. If you remember geometry, it might be obvious to you that the inverted vee takes less space than the flat top.

This antenna can be fed with coax, such as the popular and relatively inexpensive RG-8X, which is easy to run from the middle of the dipole to your shack. The center conductor of the coax is soldered or clamped to one leg of the dipole and the shield is attached to the other. It makes absolutely no difference which side is shield/braid and which side is the center conductor of the coax.

There are several commercially available center insulators that allow you to simply screw your coax onto the insulator. This makes it even easier since many vendors sell lengths of coax cable with the connectors already attached to both ends. Note that it is a good idea to take the strain off the coax connector at the antenna end if you can. You do not want the weight of the cable hanging down to pull the connector loose. Some of those commercially-available center conductors have a scheme for doing this. I also refer you once again to the ARRL's *Antenna Book* for some other suggestions.

Copper wire is usually used for the antenna wire for a number of reasons. The gauge (size) of the wire is not that important so long as it is big enough to adequately support the antenna but not so big that it becomes too heavy and droops. The support ropes should be weather and UV resistant unless you enjoy reattaching them often and having to toss the support ropes back over the limb again and again.

Many vendors sell rope specifically designed or especially good for this purpose. You may even find weather/UV-resistant rope at your local camping store. It does need to be light enough so you can easily throw it over a tree limb yet heavy-duty enough to hold those three insulators and a length of wire that might range from a few feet to almost 300 feet, depending on the frequency range of your dipole.

As with most antennas, the higher in the air you can get a dipole—away from the ground and other objects—the better, and especially if you want to use the antenna to work distant stations. You will make contacts, though, if it is just above head high, and in some cases it actually works better over closer range if it is low rather than if it was in the clouds.

The overall length is determined by dividing the number 468 by the frequency in megahertz. Results are in feet.

That means a dipole cut for 3.8 megahertz—near the middle of the 80-meter Amateur Radio band—will be about 123 feet long in total. Or each leg (the piece of wire on either side of the center insulator) will be about 61 feet 6 inches.

That also means that you would need supports (trees, fence post, mast, eave of the house…something relatively solid) about 130 feet apart with no obstacles between. It is a little known secret, though, that you can bend the legs if you really need to. Just use ropes and insulators to keep the wire from touching a tree or other object that you might use at the bends.

PROS: Cheap, easy to put up, works well on the band for which it is cut, and if it falls down, just put it back up. If it breaks, splice it and put it back up. You can bend the legs to fit on your lot, too. It is also relatively stealthy since it is difficult to see among trees. You could probably fit one for 30 meters or higher in an attic or beneath an eave on the house.

CONS: Needs to be high in the air for DX, is directional to some extent but with little or no gain on its fundamental frequency, and will only be close to resonant on odd multiple harmonics. That means your 3.8 megahertz antenna will probably only be useable on that band without a wide-range antenna tuner. A dipole cut for 7.1 megahertz would work okay on the high end of 15 meters but would be problematic on other bands.

Remember, coax feedline can have lots of loss in high SWR conditions, so even if your tuner makes it work, you may have noticeable loss of power if you try to use your dipole on frequencies too far away from resonance. Which brings us to fave five antenna number two.

#2 – The doublet with parallel feedline. An effective radiator since the beginning of the hobby, this antenna is really just a dipole, as described above, but instead of coax cable it is fed with open-wire feedline, ladder line, or window line—feedline in which the two conductors are kept the same distance apart from antenna to shack. Since the dipole is a "balanced" radiator and parallel feedline is a pair of parallel conductors, they really like each other. (See my window-line doublet article at my web site, **www.n4kc.com**).

Many vendors offer such balanced feedline or you can even make your own. Most of us go with so-called window line because it is strong and relatively easy to work with. It is also cheaper than coax in many cases.

If you decide to make yours—it is just two lengths of copper wire, after all—you can find several vendors who manufacture non-conductive spacers to help keep the wires equi-distant from each other for whatever distance it is from the radio to the antenna feed point. This spacing of the wires can be anywhere from a half inch to six inches, and that is what determines the impedance of the feedline.

PROS: In addition to the pros mentioned for the dipole, this antenna also works well on most bands above the one for which you cut it. Since the parallel feedline typically has very little loss even when the SWR is high, the antenna becomes a good multi-band—NOT necessarily all-band!—antenna when fed with this type line and used with a wide-range antenna tuner. A balanced tuner is even better.

CONS: Open wire feedline has some quirks that you have to take into consideration. For one, the feedline must be kept at least a few inches away from metal or other conductors, including the ground. That makes it problematic running the stuff into some shacks alongside coax or near gutters.

You will also notice that the antenna connector on the back of your beautiful new radio is a coax connector, not a couple of posts for the two wires of the balanced feedline. Most modern radios have 50-ohm unbalanced outputs. You will need a balun to make the transition from balanced antenna and feedline to unbalanced radio.

Read up on baluns when you get a chance. They can make life easier or more difficult, depending on whether or not you understand them and how they work. For right now, though, don't worry about that. Just assume that if you use balanced feedline in any of its various forms on any of the antennas we mention in this chapter, you will need a "current" balun with a ratio of 1:1 or 4:1. Many vendors offer such a device with prices from a few dollars to over a hundred, depending on what kind of power you plan on running through them. Get a mid-range one and you are likely set from now on, including later when you get adventurous.

The balun will have a coax connector on one end—the "output"—and a pair of binding posts for your balanced feedline to attach to the other.

By the way, if you have an "antenna tuner," you may already have a built-in balun. Many of the commercially available units do. You can tell by reading the manual or just check the back of the box. If there is a built-in balun, there will be a pair of connection posts back there and it may even be labled "Balanced output."

One more consideration for the dipole (or any other antenna fed with open wire or other balanced feedline): the length of your feedline is also a factor in how the antenna tunes and you may need to experiment to get the correct length for best results on the most bands.

Do not let that deter you, though. With a decent antenna matching device, your dipole/balanced feedline antenna will be very forgiving and will give you a good experience on many bands.

What? You don't think you have trees in the right place or a yard big enough for a nice dipole? Then consider fave five aerial number three.

#3 – The quarter-wavelength vertical. A vertical radiator has several advantages over horizontal antennas, such as dipoles. For one thing they are omni-directional, both listening and transmitting in all directions at once. That is handy if you do not know what direction your desired contact is in.

The vertical antenna also has a low-angle of radiation. Trust me, this is a good thing for working the most distant stations. For those of you who have to operate from postage-stamp city lots, they also require minimal space. And such an aerial can be as simple as hanging a piece of zip cord from a tree limb.

Of course, that zip cord wire needs to be insulated from the tree and run some distance away from any other metallic object like an aluminum mast, tower, or supporting structure. The formula for the vertical antenna's length is 234 divided by the desired frequency in megahertz. That's right. This gives you a result that is exactly half as much as the half-wave dipole above. That means, then, that a vertical is a quarter-wave antenna, right?

Not exactly. It still needs to be an electrical half-wavelength to be resonant on the desired band. More on that in a moment.

For a simple vertical antenna for the 40-meter Amateur Radio band, the vertical radiating element (wire, zip cord, aluminum, a flag pole, anything electrically conductive to which you can hook the center conductor from a piece of coax) is only about 33 feet long. The length (height) gets to be problematic for 60, 80 and 160 meters, though. That is a shame, too, because verticals are quite effective for the higher wavelengths/lower frequencies. That is why broadcast AM radio stations all use vertical antennas. Those towers you may have noticed out back of the radio station transmitter buildings? They ARE the antennas, not necessarily the supports for antennas.

But back to our 40-meter vertical antenna. Got a tree limb 35 feet off the ground? Tie a knot in one end of a 33-foot-long piece of copper wire, run a rope through the knot and tie it tightly. Then throw the other end of the rope over the limb. Pull the wire up until the lower end of the wire is three to six inches from the ground but the high end is not touching the tree limb at the top. Tie off to the tree trunk the end of the rope you are holding.

Now we have to pay the price for the vertical element being so short. We have to provide a ground field against which the vertical "works." This ground field is the other half of the antenna.

To build the field, lay out pre-cut pieces of wire that are approximately 33 feet long, stretching each one outward from where the bottom of the vertical wire hangs—but NOT connected to the vertical wire—and array them in a radial "spoke" pattern, stretching them out from the middle, just below where the end of the vertical wire dangles.

How many pieces of wire? Use 20 or 30, which we will call "radials." Are 20 or 30 enough? Can you get by with 4 or 8? Yes and yes. But getting at least 20 is a good thing and then, if you add more, it is even better but you begin to get diminishing returns. Getting more short radials is also better than fewer long ones if you do not have 33 feet with which to work in all directions. Tie the radials together where they all come together in the middle, beneath the vertical, but again, do not attach the bundle of radials—now all electrically connected—to the vertical wire.

What you should do is attach by soldering or solidly clamping the shield (braid) side of some RG-8X coax to where the radials are twisted together. Solder or clamp the center conductor of the piece of coax to the bottom of the hanging wire. Weatherproof it as best you can and tape it up so the coax shield/bundle of radial wire ends cannot accidentally make contact with the coax center conductor/vertical wire hanging from the tree limb.

If you want to go ahead and give the antenna a try now, you can. Just run the coax into the shack, hook it up, and see how it plays. If all works well, there are a few steps yet to help make the installation more permanent.

Trench out a shallow ditch and bury the coax (make sure that it is approved for burial beneath the ground) to a point where you run it into your shack. You can also bury the radial wires if they appear to be a threat to become entangled with a lawn mower blade, a happenstance that can cause great consternation and cussing. In most cases, though, you can use simple lawn staples to tack them down and they will soon disappear beneath the thatch in most lawns. Some vendors actually sell staples for this very purpose.

Safety note: RF current does run along those radials, and the ends farthest from the vertical wire can be especially "hot." Do try to keep them as low and out of sight as possible. Should they be insulated? Not necessarily, though that does lower the chance someone will touch one and feel a tiny bite, and it can also make the copper last longer if the soil is especially corrosive for such metals.

There are several commercially made verticals that offer more strength and, through the use of traps or other technology, make them multi-banded. I use a Hustler 4BTV and it is a good antenna. I bought it used for $50. But note that you still need a radial field under any quarter-wavelength-long vertical antenna that is ground-mounted, no matter what the sales pitch says.

By the way, a vertical can be mounted above ground, usually on a mast, and this setup has its own benefits as well. You still need at least two radials for each band on which you will operate. Cut those to be a quarter wavelength for each band. You may even need to do some tuning on those radials, cutting or lengthening them to get the lowest SWR on a particular band. That is why they are usually called "tuned" radials. Those buried beneath a ground-mounted or near-ground-mounted vertical are not quite so persnickety.

Even so, do not worry about 1.5- or 1.7-to-1 SWR and waste hours of perfectly good operating time trying to get it to a flat 1-to-1! Anything 2:1 or better is fine.

If you have a multi-band vertical mounted above the ground on a mast or other support, you should have a couple of radials tuned for each band that you use. For the same reason a 40-meter dipole will work on 15 meters, a 40-meter tuned horizontal radial will also work for 15 meters.

You will sometimes see such an up-in-air vertical referred to as a "ground plane." That is because the tuned radials beneath it form a ground plane against which the vertical element works.

PROS: A simple and effective antenna, the vertical antenna is omni-directional. Since it is vertical, it takes very little space to erect, though putting down radials could be problematic. Many Hams raise them when they want to operate and lower them when not on the air. This means they can be great portable antennas if you are operating from an island or mountaintop or campground somewhere.

The angle of radiation of a vertical antenna is most conducive to working longer distances.

Commercially made multi-band verticals are available from many manufacturers and retail outlets and can be quite inexpensive.

CONS: It is an omni-directional antenna, so you reel in signals from and cast out RF into all directions, not just the one in which the station you want to work happens to be. That means you are also hearing signals from directions in which you have no interest and are dividing up the power you are sending out into a perfect circle instead of concentrating it in one or two directions.

A vertical antenna is also more susceptible to manmade and electrical noise than a horizontal antenna. That is why it has a reputation of being "noisier" when receiving.

The antenna requires a radial field…the more wire the better…and you may not have enough real estate to stretch out radials in all directions that are equal to the height of the vertical radiator. That can be quite a bit of wire! The feedline also needs to be buried for a ground-mounted vertical, preferably below the level of the radials, or it can pick up stray RF and ferry it right into your house.

Even so, if you can get any radials down, regardless of whether or not they are uniformly spread out beneath the antenna or some are not quite long enough, it will still work.

#4 – The horizontal loop. One of my personal favorites, the horizontal loop can be a good performer, stealthy, and will fit on many smaller lots that a full-size, half-wavelength dipole won't. (See the loop article on my web site at **www.n4kc.com**. This has become by far my most requested-for-reprint and downloaded article.)

A loop is just what it sounds like: a big loop of wire, supported by anything you can find to hold it up it as it makes its way around the backyard or the entire lot. Many Hams tack the wire beneath the eaves of their house all the way around. Others erect poles or masts at four corners and make a square loop.

When you bring the ends of the wire together at the feed point after making that big loop around the yard, you use a short insulator to tie them together, leaving a small gap. One conductor of the feedline is attached to one end, the other side to the other end. You can feed with coax or open wire feedline, but open wire/ladder line/window line is a much better choice if you want to use the antenna on bands where it is not resonant.

Good news: the loop will be resonant on all harmonic frequencies, not just the odd ones. That means a loop cut for 7.1 megahertz will be close to resonant on 14.2, 21.3, and 28.4 megahertz.

For a resonant loop, the wire length should be calculated using the formula of 1005 divided by the desired resonant frequency in megahertz. A loop cut for 3.8 megahertz is about 264 feet long. A perfect loop is arrayed in a circle, but a square, diamond or rectangle shape is fine, so long as the rectangle is not too "skinny." If you plan to use the loop on multiple bands, simply make sure it is cut at least as long as needed for the lowest frequency on which you intend to operate.

Like the dipoles, a loop performs for DX better if it is higher in the air. However, on some bands, it will still be a decent low-angle radiator at relatively lower heights.

PROS: This is one of the quieter choices for an antenna. It virtually ignores much manmade noise so is a good choice in urban areas where the airwaves are polluted by wall warts, computer power supplies, and power line grunge.

A horizontal loop can often fit on real estate that a dipole will not, especially considering its shape does not necessarily have to be round. It can be supported by whatever trees or other structures you happen to have. You do not have to rely on trees or other supports being strategically placed on your lot. And if you use insulated wire, you can run it right through bunches of leafy tree limbs and bushes with little adverse effect.

As noted, the antenna may be long but it is also very stealthy. The wire is almost invisible among trees and even if in the open, it virtually disappears in the sky unless you use really heavy gauge wire. The feedline will be the most likely part that is visible, but even that gets pretty difficult to discern on most lots.

Here is a very positive benefit. The loop also has useable gain, especially above the fundamental frequency. If you look at the radiation pattern for a horizontal loop, you will see lobes indicating a stronger signal in certain directions. Those gain lobes increase in number as you go higher in frequency from the band for which the antenna originally cut, the one where it is a full wavelength long. Use open wire or ladder line to feed the loop, employ a wide-range antenna tuner, and it becomes a very good multi-band antenna.

Many often ask about making the antenna two or more wavelengths long. More wire, better antenna, right? Some have the real estate and the wire and want to give it a whirl. Most who have experimented report improved performance at two-wavelengths, but not dramatically so. There is a point of diminishing returns.

That plus some of the cons below are even more of a problem with such a big antenna.

CONS: That much wire can be heavy, causing it to droop. It also requires maintenance since lots of things can happen to a stretch of wire that long (A 160 meter full-wavelength loop is almost 560 feet long! A two-wavelength loop for 160 is almost a quarter of a mile of wire!).

It stands to reason, too, that if the antenna can capture distant signals, it can certainly snag lightning, too. Blowing dust, rain or snow can also create a lethal voltage static charge at the end of your feedline. Make sure the antenna is grounded—or at least disconnected outside—in such weather.

With those gain lobes mentioned above, you also get nulls. If the station you want to work is in the middle of a nice lobe, super! If he is in one of those deep nulls, "Sorry, old man, you're down in the mud!"

Note, too, that if you go with a two-wavelength version, those nulls become problematic on more bands while the gain lobes get skinnier more quickly.

#5 – The G5RV. The most discussed, maligned and misunderstood of all the simple antennas! Introduced by a British Ham with the call sign G5RV, it has gotten a bad rap because so many manufacturers claim it to be an all-band antenna, "using only your rig's internal tuner!" Well, no.

By generally accepted definition today, the G5RV is a 102-foot-long dipole, fed with a matching section of 450-ohm window line, and then it uses coax the rest of the way to the shack. My authority on this is none other than the ARRL's *Antenna Book*. Take it up with them if you disagree with this description.

In my experience, the G5RV will work fine on 40 and 20 meters, is not bad on 60, 17, 12 or 10, and might work okay on a narrow portion of 80/75. I have one—and I like it a lot—and can tune it with a good tuner on a few other bands. Still when I get beyond those bands mentioned above using a wider-range "antenna tuner," it is mediocre at best and the internal "tuner" in my rig will not even come close to finding a match. It merely fusses and refuses to try on those bands. (In all fairness, my G5RV is a derivative version that is shorter than the classic version.)

Google "G5RV" for several construction articles. There are also G5RVs available that are commercially made but I cannot vouch for any of them. I would be leery, though, of any that claim you will be able to have fun on "all bands with your rig's internal tuner!" Check the reviews for various vendors' versions on eHam.net. Most use the same basic design but some are more strongly constructed. If you buy commercially, stick to the ones with the higher reliability reports. They will all work but not so well if last night's thunderstorm put it on the ground.

To get on the air the same day I bought my transceiver several years ago, and to be able to effectively access as many bands as possible, I purchased a kit from The Wireman, a vendor you can find using a quick web search. I actually paid less than the wire, ladder line and insulators would have cost me if bought separately. That G5RV has been up ever since and is sometimes my best antenna on 40.

Ignore those who denigrate this fine compromise antenna. But also ignore those charlatans who claim this to be an "all-band antenna." A wet shoestring is an "all-band antenna" if you can somehow make your antenna "tuning" device find a match. Just not a very good one!

But the G5RV is a very good multi-band antenna choice for getting on the air quickly, easily and effectively.

PROS: The G5RV is a good antenna on several Ham bands, yet it is shorter than a dipole for 80/75. On 20, it produces four gain lobes in a cloverleaf pattern, which gives you a very good signal in those four directions.

It is cheap and, if you follow measuring instructions precisely for both sides of the dipole as well as the ladder line matching stub, it is easy to build and hang.

The G5RV can be used in an inverted vee configuration, too, if you lack the room or end supports.

CONS: It is not an all-band antenna, any more than any length of random wire and feedline is an all-band antenna. You could create some problematic mismatches, even for a good tuner, if you do not follow recommended measurements closely. The window line should hang down vertically as far as possible so you really need to be able to get the feed point up at least 45 feet or so above the ground.

Remember, too, that since a portion of the feedline is coax, it will be lossy if you have a high SWR. Keep the run of coax as short as possible.

Well, you ask, could I not just run open wire feedline all the way from the feed point to the shack? Sure you can. The only thing is that this makes this no longer the traditionally-described G5RV. It is simply a dipole fed with balanced feedline, just like choice #2 above, with the same pros and cons as that venerable antenna. The advantage of the G5RV is that the matching stub—the length of balanced feedline—causes the antenna to be resonant on more than a couple of bands. That is important if you use coax for any portion of the feedline.

So there they are. As mentioned, the main reason for this exercise is to give the new Ham or someone who is returning to the hobby a bit of inspiration and some truly viable choices to consider for an antenna. At the same time, my objective is to urge them not to be too ambitious—ambitious to the point they never get around to putting anything up! And eventually just relegating that nice Ham Radio station to a corner of the basement.

I think many of you will find a great deal of satisfaction in building your own aerial and then seeing how it works. There is certainly more fulfillment in that for many of us than there is in buying something already built and simply draping it over something in the backyard.

I do encourage you to play with your design. Try things. If they help, keep them. If they make your antenna a dummy load, toss it and start over. But most of all, have fun.

But if you want to purchase an already-built antenna, that is perfectly all right, too. If the experimentation bug eventually bites you, great. If not, that store-bought radiator will give you many hours of on-air satisfaction.

Most importantly, do not use the antenna as an excuse to put off getting on the air. As you see, there are several choices you can explore, and none are really all that daunting.

Come on in! The airwaves are fine!

Here are some information sources on the Internet for these basic antennas:

- Wikipedia (good article for beginner): **http://en.wikipedia.org/wiki/Dipole_antenna**

- *QST*, June 1991 article on the dipole titled "Antenna Here is a Dipole" by James Healy NJ2L Go to **http://www.arrl.org/arrl-periodicals-archive-search** and search by author, title, or month/year. Note that this article is only available to ARRL members.

- Athens Amateur Radio club article by KV5R on ladder line and its myths and uses: **http://athensarc.org/ladder.asp**

- Short article about vertical antennas by KB6NU and followed by many useful reader comments about the article as well: **http://www.kb6nu.com/choosing-an-hf-vertical/**

- Construction guide for the G5RV antenna by G5RV himself: **http://www.astrosurf.com/luxorion/qsl-g5rv-2.htm**

Additionally, I have written a number of antenna-related articles that go well beyond the scope of this chapter. Most are specifically intended for the beginner and might be of help to you as you decide what antenna you want, how you will feed it from the radio to the aerial, and more. Here are the links to some of those articles:

Antenna feedline choices:
http://www.donkeith.com/n4kc/article.php?p=13

N4KC's horizontal loop article:
http://www.donkeith.com/n4kc/article.php?p=12

A discussion of antenna resonance and standing wave ratio (SWR):
http://www.donkeith.com/n4kc/article.php?p=32

N4KC's eHam article on ladder line doublets:
http://www.eham.net/articles/16690

Chapter Six: What to expect on the Amateur Radio HF bands

I am going to make a big assumption here. Now that you have read this far, you are not nearly as afraid of getting on the air as you were, you are well into either planning your antenna installation or already have one erected and are ready to go, and there are no other obstacles in your path to Amateur Radio bliss. If so, then it is now time to give you a quick look at what you might expect from each of the Amateur Radio frequency bands you may be considering trying out.

Now if you have been licensed since spark gap, or if you know how to sweet-talk rare DX stations out of otherwise dead bands, this chapter is not necessarily for you. Or if you have decided on one particular band already and believe the others are not necessarily of interest, or they are not accessible with your radio or your antenna, then this little bit of the book may not appeal to you.

But, heck! Read through what I have to say anyway. It might just get you off that band where you've been molting since the end of WWII or you might find you would love it when you see what else the big, beautiful high-frequency radio spectrum has to offer!

That being said, if you are new to HF—the Amateur Radio bands between 160 meters and 10 meters—let me highly recommend that you listen to each of them. Listen! And then listen some more. If you do not get anything else out of this chapter, let that be the message you retain.

Listening is a great way to learn the nature of the various bands as well as the customs among the natives that inhabit them. They do vary band to band. Even sometimes from frequency slice to frequency slice.

Now, let's start our tour at the top (if we are talking wavelength)—the "Top Band"—or the "bottom" (if we are talking about frequency).

160 meters - 1.8 to 2.0 megahertz: An interesting bit of radio spectrum, this. Compare it to the commercial AM broadcast band, which lies just below it in frequency. Note, though, that a smaller change in frequency results in not only a decided change in how this band behaves but also requires you to re-tune your radio and/or amplifier more often.

That is because even a small change in frequency results in a much larger change in wavelength way "down" there. Or "up" there. And that is also why an antenna that works just fine at 1.810 mhz may be hard to tune at 1.990.

That means 160 meters has its technical challenges. A half-wave dipole should be about 270 feet long to be efficient here, and that is more space between supports than many Hams can muster. Other antenna types can do well here, and especially vertical radiators. After all, as mentioned, just below the 160 meter Ham band is the commercial AM band. Every one of those guys uses vertical antennas. But even a quarter-wave vertical needs to be about 135 feet tall unless you employ wizardry like coils and traps. And remember, AM broadcasters only need to make their antennas/towers resonant on one frequency because they never move. You, however, may be zooming from one end of this band to the other, and that requires constant retuning.

As far as other antenna choices go, many swear by the inverted-L, which offers the advantages of both a vertical and horizontal radiator in far less space. The inverted-L features as long a vertical element as you can manage and then the wire takes off horizontally to a support somewhere. It also requires a radial field beneath it to work well because of that vertical component and the need for a counterpoise. Nothing comes for free at this wavelength. (See the section on verticals in the previous chapter.)

I use a full-wavelength loop, which rings the backyard. It does well on paths out to 750 miles or so, and I have worked Europe many times with it, but it would really need to be higher in the air for real DX. But notice that a full-wavelength loop—about 570 feet of wire—can often fit on a lot that would never hold a 270-foot-long dipole, even with the wire elements bent around trees.

Trying to use an electrically too-short antenna on 160 or 80 meters can be a real challenge. A short antenna at this wavelength can result in a very low impedance for your matching network to try to swallow. Low impedances can result in high currents and a bunch of loss of RF as heat, so if the components in your "antenna tuner" are not big enough, you could see smoke. That is not a good thing.

Is all that antenna effort worth it? That is ultimately up to you and how much you like this band.

You can pretty much forget 160 during daylight hours except for local communications of less than 25 miles or so. It is also quite a challenge to use the band in the summer in most places because of its susceptibility to atmospheric static.

However, when the weather cools, it can be a delightful band for both ragchewing and working DX when the sun goes down. From early evening on until just after local sunrise, the wintertime band is remarkably quiet and pleasant to listen to. The band is less crowded most of the time so there is room to stretch your legs. Since everybody else you run into there has more or less overcome the challenges of operating on "Top Band," there is something of a fraternal feel here.

Also, because of those challenges, contests are especially fulfilling for those who do well in them. They are typically scheduled in fall and winter for obvious reasons, and they can be a true test of not only your station's capabilities but your operating skills as you take advantage of the band's quirks. Or not.

Regardless the time of year, most stations operating on 160 meters use more power than the typical 100-watt transceiver can produce. That means most Hams on this band employ an external amplifier. There is QRP activity but not much. There is also a great deal of CW and digital use since Morse, RTTY, PSK31, and similar modes work better in marginal conditions than SSB

You may ask, "Well, the AM broadcast band is just below this slice of spectrum and I hear stations from all over the country, especially at night, so why can't I talk that far from my station?"

Two reasons. First, even the lowest-powered AM stations typically run at least 250 watts but most of those you hear booming in are putting out 50,000 watts. The most you can legally run is 1,500 watts. Secondly, those guys also have big vertical antennas with a solid radial field beneath, as mandated by the Federal Communications Commission. If you could run 50,000 watts to a 600-foot vertical with over a hundred radials, you could do just as well. You can't, of course.

Here's another thought: those commercial AM stations do not have to hear each other either. You, on the other hand, do have to hear the station with which you are communicating. That means he, too, needs to be running some considerable power to a decent antenna for you to be able to converse with each other.

As you listen to what is happening on 160, you will notice many long-established roundtable QSOs among friends on this band. Some are friendly and welcoming to strangers who are bold enough to break in to their groups. Others will either ignore you or give you an earful that might hurt your precious feelings.

I would suggest that you treat a roundtable or QSO on the air on any band much as you would a conversation on the street. See if they seem welcoming of interruption. If not, go somewhere else. There are myriad frequencies available. Don't interrupt the flow of the conversation, either. Wait for a lull. Observe these caveats and most existing roundtables are more than happy to welcome a newcomer and offer up some signal and audio reports. Even so, don't be surprised if you are told to go pound sand.

One other ugly truth: on this band and a couple of others, you may encounter some rough language and questionable subjects under discussion. My suggestion? The largest knob on most radios is the one with which you change frequencies. Use it and move on to more pleasant listening.

All legal modes of emission are allowed anywhere on 160 meters. That means you could, if you wanted to, fire up on SSB at 1805.

Please don't. There is a long-accepted band plan in effect on each of the Amateur Radio bands. Follow it, even if it is not an FCC-mandated legal requirement. It will work much better for all involved.

First, it prevents chaos. Secondly, it is the polite thing to do. And finally, you could probably bellow, "CQ," for hours on that frequency and not have anyone answer you since it is in the middle of the portion of the band reserved by mutual agreement for stations using digital modes. Oh, you may get a few responses from a few guys telling you where you can shove your microphone, and that certainly would confirm that your rig and antenna are working, but there are far better ways to accomplish that.

What is this "band plan" or "mutual agreement" thing? You didn't see any questions about that in the General exam pool? Visit the American Radio Relay League web site (**www.arrl.org**), find the chart that shows the band plan. Print out a copy. Then stick to it.

Do it for the reasons quoted above. But do it mainly because it is the right thing to do.

80 meters - 3.5 to 4.0 megahertz: "80 meters" is really TWO bands— 80 meters and 75 meters—that just happen to be connected in the middle. On a frequency-vs.-wavelength basis, it is one of our biggest chunks of real estate, and that can cause some problems if your antenna system is not broad-banded enough or you do not use some kind of serious matching device. It is somewhat of a challenge if you want to work both ends of the band with the same antenna.

Like 160, 80/75 is primarily for chats over shorter distances during the daytime—maybe out to several hundred miles—and susceptible to lightning static noise in the spring and summer months.

This band, too, offers up good sky wave propagation at night during some parts of the year. You can literally work the world if you are patient enough and have a good enough setup for a station. You will find most operators using dipoles or variations, but the vertical is a good DX antenna here, too, because of its low angle of radiation. Some stations employ several verticals, all fed in and out of phase, to create a vertical beam, and these can be very effective for DX work. They can also be quite complicated to construct and require a good deal of acreage. Verticals typically won't work as well for closer communicating due to that low angle of radiation.

There are even some beams in use 80/75. Do a Google search and you will see photos of some true monsters. You may have to employ a '68 Buick straight-eight engine to rotate one of those bad boys, though!

Also like 160, you will hear plenty of good-old-boy roundtables as well as nets, mostly because the band supports regional communications so well. The band has also gotten a reputation for having more than its share of curmudgeons, characters, and just plain goofballs, many of whom use language more appropriate for a billiard parlor. I hear them and you will, too, even though such foolishness is not as rampant as some folks seem to think it is.

You might also be enjoying a nice chat with someone when suddenly a station will pop in with something like, "Hey, this is our frequency and we are about to fire up the Goofballs group. Y'all move somewhere else." And the suggestion may not be all that courteous. Or even in the form of a suggestion.

Well, you, me and they all know they don't own the frequency, even if they have been meeting on that very same spot on the dial every night since 1957.

You could remind them of that fact and politely suggest they meet down or up five kilohertz this particular evening. You would be perfectly within your rights to do so. My suggestion is to politely thank them for letting you know about their fine group and ask your contact to move up or down to allow the Goofballs to do their thing.

Who knows? Their VFOs may have become permanently welded by now to that spot on the band!

Again, that is one reason why our fancy, modern radios are equipped with tuning knobs. Use it. If they want to show their ignorance and lack of proper upbringing, allow them to do so without benefit of comment from you. That only encourages such silly behavior.

Nets serve a useful purpose and such operation may or may not appeal to you. They can offer you a chance to serve the public interest through message handling or during emergency situations. They have a solid social benefit, too, and can be a great way to meet other Amateurs in your city, region or state.

Give the nets in your area a listen and decide for yourself if you want to join. Avoid causing them interference. Chances are, they will be polite in informing you that you are on the frequency that they use each day. It is easier for you and your contact to move up or down a few kilohertz than for a couple hundred of their guys to QSY (change frequencies).

If you hold a General class license, you may be inspired to go for the Extra class when you see the big chunk of SSB spectrum those license holders get on 80/75. Extras can do the voice thing all the way from 3.6 megahertz to the top of the band, 4.0. And the band plan calls for a DX window from 3.790 to 3.800, frequencies on which a General can only listen and salivate when Europe or Oceania stations are rolling in. There are also some great CW DX opportunities on the low end of the band, but you need the Extra to venture below 3.525 where many of them hang out.

(At the time of the publication of this book, there is a petition before the FCC to allow digital operations in the 3.600-to-3.650 range since those modes are not cramped in a narrow section below 3.600. Extra class SSB would no longer be allowed there, but that still would leave a huge voice portion of the band.)

60 meters - 5.330 to 5.405 (or thereabouts): Now this is an odd little band for a number of reasons!

First of all, it is channelized, the only Amateur Radio band that employs such a thing. You are limited to 2.8 kilohertz of bandwidth centered on any one of those five channels. If you use SSB, you can only legally use upper sideband (USB). And you can only run 50 watts "maximum effective radiated power relative to a half-wave dipole."

Huh?

This is all because we Hams share this band with another service and are not allowed to interfere with them. We are secondary to them, but though I use the band often, I have never heard anyone else but Hams operating there. There is some talk of loosening some of these odd restrictions. We will see.

Many older commercially available radios will not transmit on 60 meters. There are now many newer radios that treat 60 just like any other band but allow for the channelization. Most newer radios allow the operator to assign the five 60-meter channels as memory channels so you only have to choose which one you want and have at it.

You will not find very many antennas for sale that are made specifically for this band. As far as I can tell, the vertical and dipole are usually the aerials of choice.

All that being said, this is an intriguing band. It offers the best in terms of propagation of 80/75 and 40, and with such limited power, you are on equal footing with everybody else—assuming everybody else is following the letter of the law. It would seem ideal for antenna experimenters since whatever you use would not be so massive, and the results not masked by running scads of power.

60 meters can support good DX and there are enough people who enjoy ragchewing that you can usually find somebody to talk with, but some regions of the world do not offer Amateur privileges here so that lessens the total number of potential DX stations you might encounter. The operators you find here do seem to be a polite lot, maybe because of the restrictions presented by channelization and low power.

The band does lose its oomph nearly so much as 80/75 does during daylight hours. Communication out to about 1000 miles is common during evenings, but when the sun comes up, that soon drops back to 200 miles or less.

I have actually heard some of the big DXpeditions (organized groups who go set up operations for a limited time from rare DX entities) use 60 meters. They would transmit on one channel and listen on another. This has created some controversy since that meant they were effectively denying anyone else use of two-fifths of the band. I maintain it allowed a whole bunch of stations to use the band, and only occurred for a few hours, so no harm was done.

This band has a lot of potential and you, as a newcomer, can help to develop it. Enjoy!

40 meters - 7.0 to 7.3 megahertz: Many proclaim 40 to be their favorite band and there is good reason...except for one big negative. The negative I will talk about later, but even it has recently gotten much better.

No matter your Ham Radio interest, 40 meters supports it nicely.

CW? Plenty of it, and at all speeds from plodding to cricket-chirping.

40 meters is a good portion of HF spectrum for QRP, too. Stations with amazingly low power output can work the entire planet here, mostly during hours when the sun is down.

Digital modes? It's there, almost all day every day.

Ragchewing? You can almost always scare up a conversation, and in my experience, it always seems to be with someone interesting to talk with.

Nets? Beaucoups! And devoted to about anything you can think of, regardless your interests outside Amateur Radio.

Antennas are more reasonably sized here. A dipole is only about 63 feet in length. A quarter-wave vertical can be a bit over thirty feet tall with radials spread out on the ground or in the air around it that are about the same length. Even garden-home citizens can usually find enough room to erect a radiator for this band.

If you experiment with antennas, you can have a blast. Vertical phased arrays. Weird wire antennas with exotic names. Even Yagi beams become something a tad bit closer to practical for some stations to erect for this great band.

Here is another band where the Extras can roam free, too, all the way down to 7.125 on SSB. And the lower 25 kilohertz of the band is a CW DXer's dream, but only if you have the Extra class.

From late afternoons all the way through an hour or so after sunrise, Hams from all over the world can usually communicate to some degree. I have worked well over 150 countries on 40, mostly during the most sunspot-deficient years, and that was with 100 watts and either a vertical, a G5RV, or the big, previously-mentioned loop doing the radiatin'.

One caveat and then that big negative I mentioned. The SSB authorization in many parts of the world does not always match up with ours. You may hear a DX SSB station working stateside stations one after the other, but he is transmitting on, say, 7.095. Occasionally you will hear a W, N, A, or K calling him on his frequency. Bad form!

First of all, he will never hear that station because he is listening up the band, and certainly not the 2 kilohertz or 5 kilohertz normally used when DX operates "split." No, he's listening up "100" or up "150," and will usually periodically let everyone know exactly where. In most cases he will be quite specific. Even though he is transmitting on 7.095 on lower sideband, he might say, "This is DX1DX, listening at 7.190 for North and South America."

Remember what I said about listening? Anyone from the U.S. transmitting on SSB below 7.125 is operating illegally. Usually it is only because the op forgot to hit the "Split" button on his rig and not because he did not know better. At least, I hope that is the case.

(Now might be a good time to recommend that you read that portion of your radio's manual where it tells you how to operate split-frequency, listening on one frequency while answering on another. This is how many DX stations work. Rather than having the world calling him on the same frequency where he is transmitting, he listens up or down the band for responses. That may be a specific frequency—"Listening up 5" or "Listening at 7.190"—or a range of frequencies, such as, "Listening up 5 to 15." You need to know how to make your radio do that before you stomp all over the DX station or call forever at a spot where he will never hear you because that is not where he is listening.)

So back to the subject. What's not to love about 40 meters?

International shortwave broadcasts. They are big and strong and really, really annoying, and you will hear them from 7.200 all the way up to the top. We in the USA share 40 meters with shortwave broadcasters in other parts of the world, so they are there legally. (Most of them, anyway.) It is simply something you will have to deal with if you use this band.

There is good news on this front, though. A few years ago the broadcasters were required by international treaty to abandon the frequencies up to 7.2 megahertz. At least they were supposed to. A few stations—mostly heard on the West Coast—continue to ignore the marching orders to move. However, the flight of the bulk of those shortwave guys has certainly made the SSB portion of the band from 7.125 to 7.2 megahertz much more desirable. And more pleasant to listen to.

There is actually diminishing interest in international shortwave these days, mostly thanks to the Internet. Many countries are no longer in favor of funding these big, powerful stations for a steadily declining audience of listeners. That is terrible news for SWLs (shortwave listeners) but great news for those who enjoy operating on 40 meters.

We can already see the effects of this lessening of shortwave broadcasts. Hams operating above 7.200 megahertz have much less shortwave broadcast QRM (interference) to deal with these days and it is only going to get better.

30 meters - 10.1 to 10.15 megahertz: Now here's an interesting band! If you are a no-code licensee and have not gotten around to learning CW yet, 30 meters will be a bit less interesting for you. You are only allowed CW, RTTY and other data modes on this band. No SSB at all.

If you think you need to make the lights dim throughout the neighborhood when you transmit, 30 is not your cup of electrons, either. You can only legally run 200 watts PEP. Okay, I know some people ignore that rule. I was born at night but it wasn't last night! Still, most operators observe that limitation and that makes it great for those who do not own amplifiers.

Like the power and mode limitations on 60, these special rules serve to put everyone on more or less equal footing. "Big guns" on this band are the ones who have optimized their antennas, not their linear amplifiers. And maximized their operating skills, too.

Like 40 meters, this is a good DX band, best at night but often offering long-distance contacts in daylight hours. And while still a bit high in wavelength for a beam for most of us—meaning such an antenna would be very, very large—30 meters makes it possible for you to design yourself some very effective antennas for the band.

I have found it to be relatively empty during the daytime, except for RTTY, but it really comes alive at night or a few hours after and before local sunset.

I also notice quite a few ops using comparatively slow code here. I assume these are guys who simply cannot ignore the treats 30 meters offers but are not necessarily proficient in the CW. Other guys slow down for them, too.

This actually seems to be one of our more polite bands. Only in the most intense DX pile-ups have I heard colorful language used, and even then, unless your 4-year-old speaks CW, he'll never know that guy was questioning the marital status of the other guy's parents.

If you hold a General class license or above, the whole 50 kilohertz of this band is yours.

One other point: if you don't like contests and believe they are the ruin of radio-dom, no problem on 30 meters. As a "WARC band," (so named because they were authorized at a World Administrative Radio Conference a while back), most contests are forbidden here. Not by law but by gentleman's agreement. And it works. Even the busiest contest weekend finds 30 meters busier but not usually elbow-to-elbow either.

20 meters - 14.0 to 14.350 kilohertz: The king of the DX bands! Long live the king!

Regardless of where we might find ourselves in the current sunspot cycle, treasures abound on this chunk of magnificent shortwave propagation. Granted we have to work a bit harder for it at times than we do in others when the sun is covered with spots and much friendlier to DXers.

This all relates to the activity on the surface of our sun, which has profound effect on shortwave radio propagation down here on Earth. Such activity seems to ebb and flow over roughly eleven-year cycles. This is a topic beyond the scope of this book, but it is a truly interesting subject about which to learn more if you are so inclined. As a Ham, you are especially well placed to do just that!

The truth is that even with limited power and basic, wire antennas and regardless where we are in the sunspot cycle, I can have delightful conversations on 20 meters with stations in Switzerland, New Zealand, or even Poughkeepsie.

Not long ago, while goofing around, I easily worked two new countries, one in the South Pacific and one in a former Soviet Union country. That was with 400 watts and a homebrew hex-beam. I've worked lots of odd-sounding call signs with 100 watts and a G5RV or a simple, ground-mounted, multi-band vertical, too.

Power and big antennas make it much easier on 20 meters. Some Hams are still convinced you absolutely must have both. I disagree. You can do just fine with less on this great band if you are skillful and patient.

There are plenty of things to occupy your leisure time, too. Slow-scan TV, RTTY, digital modes, rag0chewing, nets, special-interest groups, contests, chasing DX, working counties, and more.

As noted, you are better off with a beam antenna and more power. Still, many QRP stations have confirmed contacts with ops in hundreds of countries. A vertical is a good choice on this band, and especially if your fancy favors DXing. You will also work many stations with G5RVs, dipoles, phased arrays, multi-element beams, loops, and even tiny magnetic loops.

Propagation-wise, 20 meters is best for DX around sunrise and sunset, and especially at gray line, when it is daylight or dusk at both ends of the circuit. But I work into Africa and the South Pacific at all hours if I am patient enough. You may listen sometimes, and especially late at night, and assume the band is dead because you hear no signals. But do not write it off. Just when you think it's dead, there's a station on some exotic isle somewhere calling CQ and he is threatening to bend your S-meter.

Like 80/75, this band has its share of malcontents and just plain rude guys with microphones. You especially need to ignore the kooks at 14.313, too. There is no sanity test for a Ham Radio license. These pitiful jerks should not detract from the fact that there is an abundance of interesting people to talk to, and especially on 20 meters.

I do not think I have ever had a boring QSO on this band, SSB, RTTY PSK31, or CW.

Call CQ (after asking if the frequency is in use, of course) and you never know who will come back to you—a rock star, a missionary in Bolivia, the owner of a major Amateur Radio manufacturer—or even a retiree in Florida!

And sometimes nobody.

It is easy to get lost in the vastness of this band, as well as the QRM if you do not have some power. People tend to answer CQs from louder stations if they envision a nice, long conversation. Keep trying, though. Chances are, regardless the size of your station, your signal is plenty loud somewhere.

If you are not into contesting, this band might be a challenge. There is likely some kind of radiosport event going on every weekend, and this is a popular band for contests. The big DX contests and the annual ARRL Sweepstakes render 20 almost unusable for much of their weekends. But the CW and SSB versions are on different weekends so you can move to the opposite mode and find the band remarkably non-busy. Or go to 60, 30, 17, or 12 where nary a contester can be heard.

Better still, jump into the middle of the contest and get your feet wet. These things are great for filling in the blanks in your Worked All States or DXCC list. And be forewarned: radiosport can be very addictive.

You will also find quite a few special event stations operating here, and especially on weekends. These are stations commemorating or celebrating something or the other, and in many cases they are showcasing Ham Radio for the Podunk Possum Fest or whatever. I enjoy making those contacts. Maybe you will, too. They can actually be educational and get you some nice wallpaper (certificates, QSL cards, and the like to hang on the shack wall) if you are so inclined.

17 meters - 18.068 to 18.168 megahertz: "The Gentleman's Band." Have you heard that expression used when describing 17 meters? There does seem to be some validity in that description. There is little of the rush-rush of 20 meters here on 17. For a DX band, there are more ragchews across oceans than on other parts of the spectrum, it appears. Even DXpeditions seem slightly more sedate when working 100 stations per hour on 17.

17 quite often offers very good propagation and does not go to sleep nearly as much as its distant cousin, 15 meters. The atmospheric noise seems quieter here than on 20, too. Antennas are more modest in required size. A mini-beam, hex-beam, or vertical performs well. And for some reason, power seems to be less an issue.

Still, sometimes when I tune the band there is nothing but hiss and a couple of birdies internal to my fine rig. Merely an hour ago, the West Coast stations were booming in with an occasional ZL (New Zealand) coming through to the southeast USA.

Sometimes you can be enjoying a nice chat with a VE7 in British Columbia, a station that is well above S-9 on the meter, and then, in a few seconds and with a "whoosh," he's gone.

Then five minutes later, he's back, stronger than ever, talking with a station in Nicaragua. Until he disappears ten minutes after that, just as suddenly lost to the vagaries of 17-meter propagation.

The band seems less crowded than others, even though it is relatively narrow. I believe this is partly because of the nature of the propagation on the band. It also may be a factor that so-called tri-band beams and many commercially-sold verticals simply don't cover 17 meters. My G5RV and my loop work beautifully here. I could easily modify my vertical, too, but why bother? I now have a hex-beam up that performs like a two-element Yagi on 17 meters. I love turning it around and hearing a station go from down-in-the-noise to in-the-room-with-me, arm-chair copy.

I am not sure why 17 meters does not get the respect or activity it deserves. It is a very interesting band that offers truly exciting propagation at times, and, as mentioned, it performs well even with relatively low power and modest antennas. Plus it is simply quieter than most of the other bands, in my own estimation. And remember, as WARC band, you will never hear radiosport here. That may or may not be a positive for you.

It has been many years since we received authorization for this band and it is surprising to me that it has been slow to become filled with eager operators. With the hex-beams and add-on vertical kits, more will surely have the ability to operate here, thus increasing usage. And with folks like me evangelizing about the virtues of 17 meters, that will cause more and more to check it out, too. Maybe I should hush!

15 meters - 21.0 to 21.450 megahertz: If 20 meters is the "king," then 15 is the once and future "king of DX." You tune there during the low point of the sunspot cycle and hear nothing but vast emptiness. Occasionally it opens to South America or the Caribbean, but mostly you will only hear "sssssssshhhhhhh." Or guys a couple thousand miles apart. But wait a few years or so. Just wait. There will eventually be signs of life!

If you like to combine ragchewing with DXing, if you enjoy working Europeans or Japanese stations on PSK31 or RTTY, if you want to close out your DXCC in one weekend, then 15 is your band when it is percolating.

Plus there is plenty of room to operate. This massive band seems to go on forever!

By the time we arrive at this point in our tour of the HF spectrum, the size of an antenna is starting to come down to the point that a full-size Yagi is possible on a postage-stamp-sized city lot. Heck, a dipole is only about eleven feet long on each side of the center insulator. There are many relatively inexpensive tri-band beams available and they all cover 15 meters (in addition to 20 and 10). And despite the sheer size of the band in hertz, it is relatively small in wavelengths, so most any antenna will be efficient from bottom to top.

It's true that when the sunspots begin to go into hiding the band is mostly "sssssssshhhhhhh." But it is truly a great ride when Old Sol blesses us with lots of spots on his face. Stick something up in the air and join in the fun.

Something? When 15 is cooking, just about anything works. A dipole can work the world. A simple vertical is a magic wand. Power is not nearly as important. Low-power signals seem to propagate just as efficiently as those of the "big guns."

You have the benefit of being able to use all modes on this band, but check the band plans for where they are typically done. It is a staple for contests. In fact, sometimes when sunspots are scarce and the band should be dead, a contest can make it come to life. That tells me the band is open more than we realize but nobody is calling.

Call and see who you can wake up.

12 meters - 24.89 to 24.99 megahertz: Now this is one odd band! It is gerrymandered in there between 10 and 15 meters, and is an even or odd harmonic to no other HF band. I find myself having to go back to the band charts sometimes to see where I can legally use which mode. Like its siblings on either side, it appears to be dead, dead, dead some of the time during the sunspot lull. Then, for no apparent reason, will suddenly spring to life. And when sunspots do start to reappear, it can pop like crazy, delivering signals from all points of the globe.

That unpredictability is one thing that makes 12 meters so endearing. Or so darn maddening. Either way, it is absolutely a fun band.

Also like 60, 30 and 17, many Hams simply are not aware of this band, older gear does not include it at all, many amplifiers don't have it available or if they do, it is an afterthought and may have input matching problems. Also like the other bands, there are no contests—ever—on this frequency range.

That means less activity. Less activity until there is a big contest underway and non-contesters flee to the safe harbor of 12 meters.

I can verify that 12 sometimes offers surprising propagation, even when Old Sol sleeps. I have worked most of the major DXpeditions over the last several years here. It is not unusual to enjoy a nice, long QSO with a VK (Australia) or JA (Japan) on 12 with no fading at all, in the middle of the afternoon.

SSB, CW, RTTY, PSK31 and other digital modes are available here and activity is growing. Like 17, it is a narrow band, but QRM seems quite manageable so far, unless there is a major DXpedition on the air.

Also like 17, as word spreads, easy antenna options become more available, fewer Hams rely on old gear that does not include the band already built-in, and conditions improve, more and more Hams will play in the nice little playground that is 12 meters.

10 meters - 28.0 to 29.7 megahertz: Want to work JAs and VKs with 100 watts and a coat hanger for an antenna? Dream of communicating across the sea with 5 watts and a whip? That is 10 meters at the top of the sunspot cycle.

"Ssssssshhhhhh." That is 10 meters when there are no sunspots.

But even then, 10 is not a total loss. This band often presents sporadic-E propagation (look it up...it is beyond the scope of this book...but it is defined in the Amateur Radio dictionary later on), regardless of sunspots, which can allow communication up to several thousand miles. That happens mostly in the spring and for a few weeks in early winter, but can occur at any time.

10 meters is also a good band for local communication up to a hundred miles (or more if you have elevation, a good antenna, and power) on a regular basis, but then, when the band does open up, it can really mess with local nets and regular roundtables.

Do check with local clubs and net directories and see if there are regular 10-meter get-togethers in your area. Just listen, too, for regular ragchews or special interest groups. I used to play chess on 10 meters with a Ham across town. Sometimes guys from the West Coast would pop in and question my moves.

This band offers the first real use of propagation beacons, with a bunch of them just below 28.3. This enables you to learn quickly if the band is open. As with some of the other bands, they may well be propagating like crazy but if nobody transmits, nobody knows it. These continuously transmitting beacon stations let you know signals are available.

It helps, by the way, if you know Morse code. That is what the beacons use to identify themselves and tell you where they are located. But even if you don't speak Morse, and you tune down there and hear them chirping away in spots where there normally are not any signals, then you know the band is open to somewhere and you can cast out a "CQ"—back up there above 28.3 or below 28.2, of course, since this portion is supposed to be reserved for the beacon stations.

Note, too, that Technician-class licensees have SSB privileges on this band. It is a great way to get a small taste of HF without knowing Morse code.

One more interesting aspect of 10 meters is that there are FM repeaters and simplex available, as well as satellite downlinks. Yep, the same kind of FM repeaters you may be accustomed to at VHF and UHF. It is a real trip to hear guys talking to each other on a repeater thousands of miles away, and you can jump right in and join them. The *ARRL Repeater Directory* (available through most Amateur Radio distributors or via the League's web site at **www.arrl.org**) and several web sites list many of them, as well as the access tones that most of them require, for obvious reasons.

You will generally hear the FM repeaters between 29.6 and 29.7 mhz and the satellite downlinks are between 29.3 and 29.5 mhz.

So there is my quick travelogue of the Amateur Radio HF bands. I think if you give each of them a try, you will discover that each has its own personality and appeal, as well as its own set of positives and negatives. Then, if you like particular ones, you can make sure you have antennas that allow you to fully experience them. Just do not dismiss any of them because of what you may have heard from other Hams, or some preconceived notion of what that slice of spectrum is like.

Oh, and do not forget that vast amount of spectrum we may use above 30 megahertz either. 6, 2, 222, 440, 900, and 1200 meters all have their own characteristics, challenges, and fun. And there is even more above those that you are free to use. Maybe not day one, when the ink is not yet dry on your Ham license, but they are there, awaiting your signal to bring them to life.

We really should be thankful that we have access to such a broad range of precious radio spectrum and that it offers us such a wide variety of conditions and signal propagation possibilities.

We as Hams received many of these bands in the beginning because regulators believed they were worthless. We proved them wrong and they became super-valuable. There may come a time when that value leads to other services wanting to take over some of it. That is especially true of the VHF and UHF portions of the airwaves, so coveted by wi-fi, cellular, and other newer communications technology. It has happened before when the 11-meter Amateur Radio band was taken away to start up the Citizens Radio Service, Citizens Band.

It will be much easier to retain what we have if we are actively using it all, experimenting with it, using it to serve the public.

Chapter 7: But what do I say...?

So you got your license, bought a rig, put up an antenna, and tuned around the bands. For many of you, that is all it takes. You are on the way to the enjoyment our wonderful hobby of Amateur Radio can offer.

For others, though, you have just hit yet another brick wall. You can legally emit radio frequency energy, but you have no idea what to say when you do. Or fully understand exactly what all you can do with that brand new license and whiz-bang radio sitting over there on the desk.

Don't worry. It is not unusual at all to feel a bit skittish about making that initial on-air contact. I dare say most of us—regardless our personality—are hesitant to jump into such a big pool. Many of us are just afraid of goofing up and having some veteran of the airways make fun of us. The truth is, the best way to get over that barrier is to burst right through it, get it over with, and have yourself some on-the-radio fun.

I do not for a second pretend to know it all. I doubt if I even know a smidgen, a word my dad used that apparently means itsy-bitsy. But after more than fifty years in this fascinating hobby, I think I have picked up a few things by osmosis. Much of that came from other Amateurs who took the time and effort to give me the benefit of their experience, sometimes calmly and patiently, sometimes not so politely or diplomatically.

For some reason, we call those kind and helpful folks "Elmers," though I have always made the case for using the term "mentor" when approaching those who may not understand such an archaic-sounding term. In the spirit of those positive mentors—and as a way of thanking them—I would like to offer some suggestions I would make to any new Amateur Radio op out there who might stumble upon this book. I truly believe that if you will take this advice, you will get more from the hobby. And, in the process, that first time you key the microphone and say those new call letters will be far less traumatic.

Now, I am not so egotistical as to think I have the secret to eternal happiness in Hamdom, or that I have all the answers. And, for that reason, here is my first suggestion:

Listen! Then, listen some more. Listen before you get your license and after it shows up in the FCC database.

I know you are itching to key that microphone or slap that key, but listen to how the other guys do it. Sure, you will hear some operators doing things that others may not think are correct. Things you *know* are wrong. You may even hear some things that will curl your hair. There are goofballs in any hobby.

In general, though, you will quickly get a good idea of procedure, accepted behavior, and standard protocol. Hopefully, you will notice that most people speak English so you will realize that you can also limit your use of the jargon that develops in any tribe. If you run across something you don't understand, make use of the dictionary that comprises a good portion of this book. That is also a good reason to keep this book next to your operating position.

Remember, Ham Radio, as previously noted, is a tribe. Mostly a very friendly and accommodating tribe.

You will also soon realize that some of those tribe members that you hear actually have some interesting things to say. I would be surprised if you do not get caught up in some of the conversations on which you eavesdrop. There are some fascinating people in our hobby! I did not promise that all will be golden, but many are.

You will also notice very little usage of terms and practices that you may have experienced if you previously operated on the Citizens Band frequencies. Remember that the FCC originally took an Amateur Radio band away in order to create that very specialized service. Some ill will lingers. Additionally, many Amateurs see CB as a bunch of Ham wannabes who either do not have the will or intelligence to pass the exam.

I do not agree with that assessment. I believe CB radio has been a de facto pathway for many to enter into Ham Radio. Just understand that there may be some irrational prejudice against you if you came from CB. Then, if you persist in using slang and operating practice that may have been just fine on 11 meters, be cognizant of the fact that you are only making it easier for those few nuts out there to pick on you.

Example? We generally give our "name" or "handle," not our "personal." Actually, "My name is Don," is perfectly fine on Amateur Radio. When in doubt, use plain language, not what you believe to be acceptable terminology to try to fit in.

When you think you have a pretty good idea of how things are done on the Amateur Radio bands, muster up your nerve and join right in. Most Amateurs welcome new ops. Especially when they don't yell, "Contact! Contact!" and ask for everybody's "first personal."

That brings me to my second suggestion for the newbie to Ham Radio:

Make a contact. Ignore sweaty palms and slight heart palpitations. Invite someone to talk with you on the air. How?

One way would be to call "CQ." CQ is just a general call that means you are looking for someone to speak with. On the HF bands, a CQ is quite simple. Using the phone modes, it can be as easy as:

"CQ. CQ. CQ. Calling any station for a contact. This is N4KC calling CQ and N4KC is listening for a call."

Remember, you are relying on someone tuning across the band and hearing your request for a QSO partner so you may have to make the call multiple times. Maybe even many, many times.

If you have listened and the band seems active, and if your gear is all working properly, you should be able to get a response before too long. If not, do not assume your radio is broken. Move to another frequency elsewhere in the band, make sure it is not in use, and try calling again.

You might also tune around and listen for others who might be calling CQ, too. You will hear them if you listen long enough. Keep in mind that the call might be restrictive in some way. If you hear a stateside station calling, "CQ DX," do not try to contact him. He is looking for someone outside the USA, a DX contact. If the station calls, "CQ Wisconsin, Alaska or Utah," he is likely trying to complete his Worked All States award. If you live in Minnesota, do not answer his call just because your state adjoins one of the ones he is seeking.

In your quest for that initial contact, you might also interrupt an ongoing conversation with the idea of joining in. As mentioned earlier in this book, you should treat a QSO already in progress just as you would a conversation on the street or in a restaurant between two or more strangers. While you may have amazing insight into the topic under discussion, you may or may not be welcomed into the chat. On the other hand, if the group seems accommodating and informal, give it a try. You may make a bunch of new friends.

A more sure—and safe—way to handle this would be to wait until the group breaks up. Or take advantage if a two-station chat seems to be coming to an end. Write down the call signs of the stations involved and, if you have them, the locations and names of the participants. Once they sign out with each other, give one of them a call and see if he or she is interested in carrying on for a bit longer.

Of course, do not call the station who just said, "Well, I am running late for an appointment," or the one who announced, "The wife just called me for the third time so…" Pick one that did not seem to be in a hurry to get off the air and on to other business.

Remember, most Hams are on the air because they enjoy meeting and speaking with other folks of a like mind. You are another "folks" and have interesting things to offer. Be confident and give the other Amateurs out there the pleasure of getting to know you.

Talk! I can tell you about many who worked hard to get their licenses, only to hesitate once they were finally bona fide. They could never bring themselves to push the microphone button.

Shy? Maybe. Scared. Almost certainly. Let me say it again. All of us were timid when we made that first over-the-air contact. My hand was shaking so that I could hardly pound out the characters on my old J-38 straight key when I had that first QSO.

Let me assure you, though, that there are many, many fine folks out there anxious to talk with you and learn more about you. I believe you will be surprised how excited that person on the other end of the conversation will be to learn he or she is your first contact.

As with any random group of people, there will be some who are more interested in hearing themselves talk than they are in getting to know you. On the other hand, most of us got into the hobby because we enjoyed meeting and learning more about other people, their lives, their locations, their interests, their opinions, and more.

Let me also urge you to have something to say. When you run out of something to discuss or if you feel the conversation may be dragging, take polite leave and look for someone else while allowing the other fellow to do the same thing. It might also be a good idea to keep a short list of go-to topics or questions nearby and refer to it if you need to push the conversation onward.

That list might include your employment, how you got interested in the hobby, what other pastimes you might pursue, and the like. There is nothing wrong with changing direction in the middle of a transmission:

"...and using a dipole I built up by myself and it seems to work just fine. By the way, I am an attorney here, specializing in business law. I have been at that job for ten years now. When I am not on the radio, my wife and I enjoy hiking up some the mountains here in the area and do some primitive camping. I'm looking forward to taking along a radio and antenna at some point and see how I can do out there."

Right there, in a few sentences, you have given the other station several key points he or she can pursue, elaborate on, or ask questions about. Note that you do not have to have unusual or quirky job or hobby points to elicit interest from your chat mate.

Of course, if nothing you have to say seems to interest the other guy or gal, or if he or she offers nothing on which you can comment, politely say, "Well, I think I will move on and see what else is happening on the band. Thanks for the nice chat and I hope to see you later."

You will not offend the other operator! That is the way it is done. Most Hams understand you want to make multiple contacts in a session. But if you want to hang around and talk at length, and the conversation is a good one, stick with it. The other op will politely bail out, too, usually simply to go make other contacts. Do not assume it is because you are a dud in the conversation department.

Start with the basics when you begin a QSO. It is customary to begin with the signal report, of course. This is actually of some importance. It is good to know that the other station can hear you well enough to continue the chat. It also lets the op know that you copy adequately, too. If the circuit is a bad one, it is best to bail and try again another day.

On the phone modes—SSB, FM—the signal report typically uses the "59" scale. The first number is a one-to-five scale for readability. How well can you understand the other station's transmissions? That scale is:

1 -- Unreadable

2 -- Barely readable, occasional words distinguishable

3 -- Readable with considerable difficulty

4 -- Readable with practically no difficulty

5 -- Perfectly readable

If you cannot understand anything the operator is saying, you have to give him a "1" for readability. Do not be afraid that you will hurt his or her feelings. It may or may not be the fault of the fine station and antenna being used. It could well be propagation that renders the circuit untenable.

Regardless, it does not matter. If the other station is a "1" then the conversation cannot continue at that time.

The second number is for signal strength and this one is a nine-point scale:

1 -- Faint signals, barely perceptible

2 -- Very weak signals

3 -- Weak signals

4 -- Fair signals

5 -- Fairly good signals

6 -- Good signals

7 -- Moderately strong signals

8 -- Strong signals

9 -- Extremely strong signals

Your first reaction is probably that this is a very detailed and dismayingly granular scale. What is the difference between "2" and "3" in the real world? To which I answer, "Don't worry about it."

If you are hearing the station pretty well, not missing a word, give him a hardy "5-9." If he is really, really weak but you can still hear just about every word, give him, say, a "4-4." Accuracy is not a big deal here.

By the way, there is nothing wrong with elaborating in plain old English, such as, "Jim, I copy you about 4-8 but the signal is fading sometimes and I'm missing a word or two."

If you are one of the adventurous ones who have learned Morse code and are making your first contact in that mode, you need to know that there is a third number in a CW (Morse) signal-report exchange. That scale is:

1 -- Sixty cycle AC, or less, very rough and broad
2 -- Very rough AC, very harsh and broad
3 -- Rough AC tone, rectified but not filtered
4 -- Rough note, some trace of filtering
5 -- Filtered rectified AC but strongly ripple-modulated
6 -- Filtered tone, definite trace of ripple modulation
7 -- Near pure tone, trace of ripple modulation
8 -- Near perfect tone, slight trace of modulation
9 -- Perfect tone, no trace of ripple or modulation of any kind

This number is supposed to correspond to the tone of the CW being received. Frankly, in this day and age, you will rarely hear a Morse code signal that is anything besides an "8" or "9." Chances are, too, that neither you nor I would know a "strongly ripple-modulated" tone if we heard one. Until you can learn about such things, I suggest that you simply give a "9." That is likely what you are hearing anyway.

So a CW report might be "599" if it is very strong, perfectly readable, and has a perfect tone.

Be aware, too, that there is a completely different scale of numbers for giving reports on digital transmissions. It is called the "RSQ" system. Look up RSQ in the dictionary in this book and you can see the numbers and what they mean.

Radiosport note: In contests that require a signal report, it is ALWAYS 59 or 599. Even if the other station is a mere whisper in the din, it is ALWAYS 59 or 599. Got it? Nobody really cares so long as the contact can be consummated, the required information exchanged, and both participants can move on to the next contact.

Now, the next basic bit of information typically imparted in a more relaxed non-contest contact—be it short or long in duration—is your name. Not "handle" or "personal." Name.

Spell it phonetically if it is difficult one to understand. Do not be too cute. "My name is Bill...'Boy, I love ladies.' Bill." Or: "Bob...'Bottles of beer.'" If you do use phonetics, not only with your name or anything else you might say, such as your call sign, stick with the NATO standard phonetic alphabet below. Feel free to print it out and keep it handy.

A	Alpha
B	Bravo
C	Charlie
D	Delta
E	Echo
F	Foxtrot
G	Golf
H	Hotel
I	India
J	Juliet
K	Kilo
L	Lima (LEE – muh)
M	Mike
N	November
O	Oscar
P	Papa (Puh – PAH)
Q	Quebec (Kee – BECK)
R	Romeo
S	Sierra
T	Tango
U	Uniform
V	Victor
W	Whiskey
X	X-ray
Y	Yankee
Z	Zulu

Quite often the other station has already pulled you up by call sign on the computer and knows your name anyway. But if it is "Charles" and you prefer "Chuck," or if you go by your middle name, tell him so. Most online sites that list Amateur operators pull the name from the FCC database, and the one you used on your license application may not be the name you prefer to use in casual conversation.

My name is "Don," which can sometimes be mistaken for "Bob" or "Ron" on the air, no matter how hard I try to enunciate clearly. For that reason, I usually say, "My name is Don. Delta Oscar November." Then I go on to the next thing most of us include in that initial transmission of a QSO.

That would be a short description of your station, including the model of your rig and your antenna.

You might say, "The rig is a Yaesu 857, running 'barefoot' and the antenna is a dipole cut for 80 meters and fed with window line." This is really all you need to say about your station at this point. You can leave off the brand and model number of your power supply, microphone and the computer you are using to log the contact.

Just rig and antenna are typically all you need to start with unless you have something unique in the shack. If you built your own microphone or have an unusual antenna setup or use wind power to run your station, this would be a good time to mention such things, though. That would then give the other person the opportunity to ask to hear more about it.

"Barefoot" means you are not using an external amplifier, by the way.

Now, listen carefully to what the other station has to say. Make notes. That helps to respond to specific points the other person makes or remember questions you want to ask. Ask questions! That is the best way to keep a conversation going, on the radio or in person, and to make it interesting. Do not make it all about yourself.

I suppose the weather is still fair game for chatting. But then move on to other more interesting things, unless you are in the midst of a blizzard or your city just broke the record for high temperature that day.

What do you do for a living or what are you most interested in at school?

What got you interested in the hobby?

What do you believe will be your favorite aspects of Ham Radio?

What other hobbies or pastimes do you pursue?

Use the old salesman's trick: find common ground. You will be surprised how often you and a Ham operator across the country (or world) have something in common that leads to a very pleasant and quite memorable chat.

Again, it is a good idea to make notes about what the other person says to help you to remember to comment on them. In short, carry on a conversation, just as you would if you were sitting in the park on a bench next to your new friend.

In a way, that is exactly what you are doing. You will also notice that if you show a true interest in the other person, he or she will reciprocate, and that makes for the best conversations, on or off the air.

If the other operator needs or wants to move on, he or she will tell you so. Do not be offended if the person ends the QSO with you and then promptly calls CQ and starts up another one with someone else. The op is not running away from you! That is the way it is done, and is no reflection on you or your ability as a raconteur.

A big part of our hobby is communicating. So communicate!

Ignore the curmudgeons! In the spirit of full disclosure, I will admit that not all the people you meet on the air are as pleasant as I have so far described them. There is a certain segment of our hobby that can be downright discouraging to all of us and especially to newcomers.

This is not limited to Amateur Radio, of course. You would find similar impolite duds if you pursued stamp collecting or quilting or fly-fishing. These are the goofballs who resent anyone else trying to get into their protected little tribe. Often that is because they fear newcomers will attempt to change something the curmudgeons think is perfect already. The one thing they fear most is change. Also many of them resent people who—they believe—got into the fraternity without having to go through the same hazing as they once did.

You will absolutely hear some ignorant Hams decry the loss of Morse code as a requirement for getting a license. Most Amateurs have long since gotten over that one, but a select few never will. The fact that you had no choice about taking a Morse exam when you took the test is lost on them completely. And never mind that CW operation has actually gone way up since it was no longer required that new licensees learn the code. Logic is not the strong suit of the curmudgeon.

You will hear other curmudgeons moan and cuss and claim the exam today is so easy it should come in a box of Cracker Jack. Why, way back when they took the test, you had to actually know some serious electronics. Today's Hams can memorize the answers and therefore do not know squat about how their store-bought radios work. Appliance operators! That's all they are!

That is more of the hazing that the curmudgeons feel should remain a part of the process. To keep out the riff-raff, they say. Never mind that by far the most and worst rule-breakers—according to no less an authority than the Federal Communications Commission—are older Hams.

Or that many of these cranks probably could not come close to passing that exam that they decry. Maybe they could change a tube in a transmitter back when but lots has changed in electronics, and some of the loudest detractors out there—the ones who abhor change—haven't even noticed.

Again, these are mostly people who are either naturally prejudiced against anybody who is not like them—young, vibrant, smart, active, new—or simply fear change of any kind. Do not allow your feelings to be hurt when—not if—you run across one of these ill-informed killjoys. As mentioned already, their presence in our hobby is in about the same proportion as the number of them in the population in general.

Let me emphasize strongly that these curmudgeons of which I speak are a tiny sub-set of the people who are active in our hobby. Yes, you will hear them on the air. Yes, if you stay around and are active enough, you might well encounter a few of them.

Do not, though, allow such boors to discourage you. And do not be timid when it comes time to make that first contact because these frequency lizards exist.

Get on the air...now! And enjoy yourself. There is a whole wide world out there waiting for you to tickle the ionosphere with radio frequency energy from your radio station.

Before you know it, you will be conversing with the best of them. Then, one magic day, you will become some other new Ham's very first on-air contact.

Chapter 8: ...and do?

Those of us who have been around for a long time sometimes forget that things that seem obvious to us may not be quite so apparent to newcomers. Maybe you have a general idea of an aspect or two of Ham Radio that seems interesting to you. That was, after all, why you went to the trouble to learn more and start working toward a license.

That aspect might have been weather-spotting. Emergency communications. DX. Software-defined radios. Building kits. Playing with antennas. All are great and worthy pursuits.

Still, as you finally begin your on-air quest, you really have little idea of all the things you can do in this truly remarkable hobby. Besides talk, that is.

I predict most of you will discover on your own all the activities Ham Radio offers. But I will also predict that some of you will get stuck in a rut and maybe even lose interest without ever knowing that there are other things going on that might be of even greater interest than what attracted you in the first place.

A quick visit to the web site of the American Radio Relay League (**www.arrl.org**) will give you a general idea of the various things available. Just to get you started, here is a short look at only a few of the things you can do now that you are a Ham.

Contests. We also call this radiosport. These are operating events in which you attempt to contact as many other stations in a specified period of time as you can. Some last a full weekend or even cover a complete year while others can be only an hour in length. You may be trying to make as many contacts as you can with as many countries, islands, mountaintops, or other explicit entities.

Some contests are serious competition. Others are more casual. All of them can be fun if that is your thing. However, no matter the size of the event or how competitive it is, you can be as serious or as relaxed as you want to be.

Ragchew. That is what Amateurs call just talking with another operator about anything and everything under the sun. As we discussed in the previous chapter, there are many interesting people on the air right this minute, whether it is 2 PM or 2 AM. And most of them would just love to chew the rag with you.

You might also join in with established groups that share a common interest. There are on-air meetings for collectors of all sorts of non-radio things, people who work in specific fields, those who are veterans of military service, and many, many more.

You may also get a kick out of contacting what are called "special event stations." As mentioned previously, these are Hams who set up and operate from fairs, local celebrations, historical commemorations, and more. You will hear them practically any Saturday morning. Give them a call and learn something new even as many of them expose Amateur Radio to the attendees at their events.

Weather-spotting. Even with today's advanced radar and weather satellites, there is no better way for authorities to track severe weather than with trained spotters equipped with reliable communications. That can be you! I have actually heard weather-spotting Amateur Radio operators save lives by reporting a tornado, allowing the National Weather Service to issue a warning before the storm was detected on radar.

Check with your local Ham club. Most work with agencies to offer training and conduct drills to prepare for this vital part of the hobby.

Emergency and public service communications. In the same vein, trained Amateurs can offer valuable assistance when emergencies occur. Again, I have listened in as Hams provided crucial after-disaster communications for agencies from the Red Cross to the Salvation Army to local governmental agencies.

Not quite so serious but also helpful is the assistance Hams give for such things as charity running events, bike rides, and more. These are also a lot of fun to participate in and offer excellent training for emergencies should you be called upon.

Google "Amateur Radio Emergency Services."

Chasing DX. Gosh, yes! This is a big one. The almost magical thrill of speaking with other Hams in distant countries has always been a strong lure to radio hobbyists. On the shortwaves—the HF portion of the Amateur Radio allocations—you can literally communicate to the far side of the planet.

Some try to earn awards for contacting as many countries as possible. Some try to earn colorful certificates for exchanging signal reports with their counterparts in a certain country or to commemorate historical events. Others simply enjoy chatting with others in exotic lands. Still other Hams are more interested in designing and experimenting with antennas to give them a stronger signal or learning about how a radio signal propagates or the effect our sun has on such things.

Operate from unusual or interesting places. A very popular aspect of Amateur Radio is taking rigs on camping or hiking trips. Many RVers have complete stations with them and enjoy making contacts from wherever they make a stop. Groups are dedicated to putting stations on the air that operate from islands, mountain summits, and even lighthouses.

Quite a few Hams work to make their radios super-portable so they fit into backpacks, dogsleds, snowmobiles, canoes, or the overhead luggage compartment on commercial airplanes, all so they can take their hobby with them wherever their travels and interests take them.

Work the satellites or the Hams aboard the International Space Station. Right this moment there are satellites orbiting the planet that were built and are maintained by Amateurs. You are invited to use them to communicate, too. You can do so with relatively simple equipment.

Most of the crewmembers of the International Space Station are licensed Hams and often take a few minutes to chat with Amateurs back on Earth. They also set up special sessions to speak by Ham Radio with students of all grade levels in schools around the world.

Of course, you can also bounce signals off the moon, off the tail of a comet, or even off the Northern Lights.

Experiment. Ham Radio can be as technical as you want it to be. You do not necessarily have to learn more than just the basics required to pass the license exam and keep your station operating within the regulations. But if you want to do more, you certainly can.

With the integration of computers and radio and the emphasis on digital communications, our hobby is just as exciting when it comes to emerging technology as it was back in its earliest days. Back when Hams were inventing a little thing called radio.

Remotely controlling a radio using the Internet is very popular nowadays. So is creating new modes that can be used to communicate when signal paths are almost non-existent. The scientists who developed super-small antennas for space travel were almost all Amateurs.

Whether your thing is computers, antennas, circuit design, or anything else on the technical side, Ham Radio will certainly be to your liking. As the "maker movement" grows so rapidly, many of the people who are fascinated with learning and using developing technology naturally gravitate to the hobby.

And this leads us into what has already been a subtle thread running throughout this chapter. This final thing I will discuss about what you can do in Amateur Radio may well be the biggest and certainly the most important. At least I think so!

You can not only have fun, offer public service support, experience the magic of international communications, compete in exciting radiosport events, build your own station, capture the best of the digital age, and accomplish so much more in Amateur Radio, but you can also:

Learn! Maybe you are one of the ones who are perfectly happy once they pass the exam to forget what little they had to learn about electronics and communications. That is fine. Many enter the hobby for other reasons than building radios and designing antennas and being able to read a Smith chart in their sleep. (Smith chart. There is something you may want to learn about already! But maybe not. Despite what some might tell you, you do not have to. But just in case you do, start by looking it up in the dictionary in this book.)

However, I would urge you to maintain a healthy curiosity about those technical aspects of the hobby. The reason? It will enhance your enjoyment of the things you really do like to do.

Example: you got your license because you like the challenge of talking to lots of countries around the world, or to try to win contests. You will be much more likely to enjoy DXing or contesting if you understand a bit about antennas and how they work. Or how to hook up a computer to your radio for rig control and keeping your contest log. Or how to tweak your SSB audio so you can cut through the clutter to work that rare DX station or run up a big score in Sweepstakes.

What happens if you have a glitch during a big contest weekend? You may be able to do a simple repair or workaround if you understand the basics of how your rig functions.

Maybe it was the opportunity to assist with weather spotting and emergency communications that got you hooked. Knowing how radios work and how signals propagate can be very useful if you are handling communication from a disaster scene and have a problem. Or in planning for a community service event. Or in putting together the necessities for backup emergency power.

Still don't think the technical side is all that interesting? Well, there are plenty of other opportunities for learning in Ham Radio, and maybe without even knowing you are doing it. When you work that station on a rare island in the Pacific, you can look it up. You may be surprised what you will learn. That special event station you worked last Saturday? It could have been commemorating the first voyage to the North Pole or the invention of the electric toaster. Before you knew it, you learned something!

In addition to all that, science has shown us time and again that we postpone many of the effects of aging by continuing to use our brains as we grow older. It is like exercising a muscle! Continuing to learn, being curious about how this stuff works, who you are talking with, where they are and more can actually keep you young.

See, right there is a benefit of getting your Ham Radio license that was never mentioned in all those books you used to study for the exam! It makes you smarter and you live longer.

Again, let me urge you to avoid getting into a rut! It is easy to do not only in a hobby but in life, work, family activities and more.

You find one facet of any hobby that you really enjoy and you do not explore anything else. You fish in the same lakes for the same kind of fish all the time. You collect the same kinds of stamps over and over. You play the same couple of golf courses week in and week out. Before you know it, you are bored with the whole thing.

Same with Ham Radio. You love chasing DX stations, working new countries, trying to get the certificates for getting a hundred of them confirmed on five bands. Great! I love chasing DX, too.

But sometimes the bands are not cooperative. Maybe I have already worked and confirmed stations in all the countries that might be available that particular day. Okay, I will just shut off the radio—though I had planned to spend all morning down there in the shack—and go watch people making duck calls on TV or something.

Working those same countries over and over? Boring!

Then, before you realize it, it has been weeks since you even turned on the rig.

Wait! There are plenty more things you can do in the hobby that may even slip up on you and become your favorite thing to do. For a while. Until you discover a new favorite thing to do. And you are never bored.

Try just playing around in a contest. Move to a different band or mode. Give digital a try. Be on the lookout for some of the special operations such as IOTA (Islands on the Air), SOTA (Summits on the Air), county hunting, or special event stations. Build a kit or try a new antenna. Write an article for one of the web sites. Try to work the museum ships when they are on the air every June, the guys who put Route 66 on the air, or the station that operates often from Indianapolis Motor Speedway. Venture outside the Ham bands and do some SWLing (shortwave listening). Use Google and find nets that might be dedicated to another one of your interests. Heck, just use that band switch and listen in for a while on a new portion of the amazing spectrum at our disposal.

Wow! That is a perfect segue to my next suggestion for something you can do in the hobby.

Combine Amateur Radio with another interest! Do not tell me you have no other interests besides Ham Radio. Of course you do!

One of the wonderful things about our hobby is that it can be easily included as a part of so many other popular activities.

Do you fly model airplanes or have an interest in starting to? They are typically "RC." Radio controlled. There you go. You are on your way already.

Are you a boater or sailor? A Ham rig in the boat only adds to the pleasure, and could even be valuable in case of an emergency. There are several on-the-air nets devoted to Hams who have rigs in their watercraft, too. Listen in on them and you may just hear an actual rescue in progress. Start with 14.300 on the 20-meter band. There are several nets on that frequency most of the day.

Camping or RVing? As previously mentioned, Ham Radio is a natural fit. Many, many Hams have stations installed in their RVs or take along low-power radios and compact antennas so they can enjoy listening or communicating from wherever their travels take them.

Hiking or more basic camping? I hear many Hams each weekend, operating from some mountaintop or wilderness area, using QRP and a simple antenna. They are, however, literally working the world from some truly wonderful spots. Go to **www.qrz.com** and look up WGØAT. Or search for his videos on YouTube. Tell me this fellow is not having a blast!

Those are only a few examples but I think you get the idea. Here is one more suggestion along the same lines. When you do combine your Ham Radio hobby with other pursuits, tell other Hams about it. Or tell fellow hikers or boaters or stamp collectors how great your radio hobby meshes with what they so much enjoy doing.

Join a club! I know. Some of us are just not "joiners." After a hard day at work, we hate to climb back into the car and drive downtown again for a club meeting. Then there are those clubs who seem to resent anyone new from diluting the perfect little organization they have built. Or they spend two hours of a two-hour meeting arguing about the club's bylaws and officer elections and never actually get around to talking about Ham Radio.

Yes, sorry to say, those kinds of clubs exist. On the other hand, there are many, many vibrant and exciting organizations out there who will welcome you heartily and will make that drive back downtown well worth it. Once you know the other members, you will know who to go to in order to get questions answered.

You may also be surprised to learn that they will volunteer to come help you hang an antenna or hook up a digital interface. They may sponsor special event stations, organize communications for community events, participate as a group in contests, and put on a heck of good time at the ARRL Field Day event each year.

The club may put on a hamfest or swap meet, too. You might want to help or just get first dibs on all that gear for sale.

Of course, they probably sponsor license classes so you can more easily upgrade to the next class. Oh, and learn some new stuff, too. Speaking of learning, many clubs give great programs on a variety of subjects related to the hobby, and that could also allow you to see other sides of Ham Radio that you had not yet considered.

At its basics, though, a good club fulfills a need we all have: to belong to a tribe, a group of people who have the same interests as we do. There is something satisfying about being around folks who speak the same language and have pursued the same pastime as you have.

Suppose, though, that despite your best efforts, you cannot find a local Amateur Radio club, or the ones near you are duds. Start your own! Talk to folks on the local repeater around the area and see if there is interest in putting together a group. Keep it informal until it takes off. Maybe make it a special-interest group. We have a great DX club in my area even though there are several good, active, general-interest clubs here. The DX club only meets once a month, has no dues, and pretty much keeps it simple.

Even if you do not join a local club, I will vociferously urge you to become a member of the American Radio Relay League. By your membership, you are making our hobby's national organization stronger and better able to protect our interests and frequencies, and to expand the scope of the hobby. You also get a great magazine, *QST*, in both print and digital versions as well as access to the archives of practically every issue ever published, and that is a wonderful storehouse of knowledge. There are many more benefits as well that make the dues a wonderful investment.

Becoming a club member—local or ARRL—also presents you with a platform to follow my next bit of advice for something you can do.

Give back to the hobby! One of the truly magical things about Amateur Radio is how so many of its followers freely give back time, money, expertise and support to the hobby.

Here are just a few ideas that occur to me. You should be able to think of many more.

Instead of selling an old piece of gear way too cheap or hauling it to the dump, loan it to a new or disabled Ham so he or she can get on the air or get more enjoyment out of the hobby. If you are able to write clearly enough, do articles for the hobby web sites about an experience you have had or something you have built. Offer to do a program at the club on that same topic. Help teach a license class. Get certified as a Volunteer Examiner so you can assist in administering exams. Support the ARRL by joining and give money to the spectrum defense effort or other causes.

There is one very big way you can give back to the hobby in which you have found so much fulfillment and enjoyment. Become a Ham Radio evangelist! Now that you are licensed and have begun to see all that the hobby has to offer, you can become a true promoter of Ham Radio.

Set up a station in the park on a Saturday and be ready to answer questions from passers-by. Be prepared to tout the hobby and your local club if someone spots your Amateur Radio license plate and asks what it is all about. Refer questioners to the club web site (if it is well done and helpful to newcomers) and to the ARRL web site. Invite friends and relatives to your shack and show them your setup. Let them hear some QSOs in progress and even send a little Morse code if you are proficient. (If you do not already, you will soon know which frequencies you should NOT allow them to hear!)

Be a mentor! Once you are comfortable doing so, offer to help someone interested in joining our ranks, or assist a new Ham in getting on the air, making that first contact, or upgrading to a higher class license. Few things are more rewarding than seeing someone you helped to get started go on to enjoy the hobby as much as you do. Or more!

Take an HT to school or work, if permitted to do so, and explain to your classmates or co-workers what it is and how it can be used in disaster situations. And just for fun, too.

It is true that some still think of Hams as that old guy, secluded in his basement, causing interference to everybody's TV on the block. Others still consider us to be nerds, or electronics nuts, with little or no social life. Still others do not know the difference between Amateur Radio and CB and think we are the guys in the Trans-Am and big-rig truck, rescuing Sally Field from Jackie Gleason.

You have the opportunity to set them straight. Tell them what we really do and what the hobby is about in the 21st century. Be ready to answer those challenges I mentioned in my introduction. Some believe we have lost our relevance, that in the day of the smart phone and Facebook, there is no reason for amateur wireless as a hobby.

Not true! We are more relevant than ever, and we still have the opportunity to lead in development and implementation of tomorrow's technology. Plus we just plain have fun doing what we do.

One of the greatest ways you can help the hobby to grow and attract other bright minds is by being an evangelist for the hobby. There is another benefit.

It will increase your own enjoyment and satisfaction as well.

Afterword: Asking for help

"I don't know how."

My granddaughter was at our kitchen table, working with her new microscope, trying to see some brine shrimp eggs, but the viewer was dark.

"You have to adjust the mirror so the light reflects through the slide," I told her.

"But Grandpop, I don't know how!"

"Did you look at the instructions? Did you try to move the mirror around and figure it out?"

"No. Fix it for me."

Ah, a teaching moment had just presented itself.

It would have been much easier for me to simply adjust the microscope so she could see the tiny eggs and then I could hurry back to my easy chair and copy of *CQ Magazine*. However, I decided to give her a quick tour of her new toy, lecture a little bit about optics and reflections, show her the section in the instruction sheet that addressed the subject, and sit her down to read it. Then I made sure she tried what she read until she got the results she sought.

Of course, she took one quick look at the infant shrimp, shrugged her shoulders, and ran off to something more flashy and glittery. But that's not the point.

The saddest words I can imagine are when someone says, "I don't know how," and then stands there, waiting for someone to do it for them.

Look, I know that not all of us are naturally and incessantly curious. Sometimes we have no desire to know how something is done or how it works. We just want it done or working. And there's nothing wrong with that. Sometimes.

I have no interest in learning how to replace the brakes on my car. I'm perfectly willing to pay somebody to do that. While the mechanic changes them out, I get to do something I want to do or that makes me money so I will be able to pay for the work. Meanwhile, he does his job, feeds his family, and the chances that the vehicle will actually stop when I want it to are ratcheted up considerably.

New kitchen cabinets? I am convinced I could study and experiment and buy a fortune's worth of tools and waste a bunch of expensive materials and learn how to build some perfectly good kitchen cabinets. But I choose to hire someone who already knows how, who already owns the tools, and who does a good job the first time. Though I do not acquire a desirable skill, I also know it is a bit of knowledge I would likely never use again. The XYL (wife) is happier, too, which is always a big, big plus.

However, when someone chooses a hobby or pastime or to get serious about any other endeavor, and he or she then makes the decision to not invest a little time and effort into learning more about it, it boggles my mind. Don't get me wrong. Because we become Amateur Radio operators does not mean we have to gain the equivalent knowledge of an electrical engineer. I do not intend to open up my transceiver and take it apart just so I can put it back together again and learn how it works. Some do. I don't. What I am saying is that we should all have more desire to learn about things in which we have interest. Why would we take the plunge and insist on asking somebody else to do it for us?

The same thing goes for many other things in life. My granddaughter, the light of my life, begged for that microscope for months. Why would she ask Grandpop to do the most basic of adjustments for her? Would you take up golf and then ask the club pro to hit the difficult shots for you? I think some folks actually would if they could!

How many people refuse to learn anything about income taxes (too complicated, no time, bad at math), then either fill out the short form because it's easier or trust somebody else to do it for them, leaving money on the table? How many people blindly invest (do not understand financial stuff, don't have time, don't want to learn) their 401K contributions in only their company's stock and lose scads of money in the process? How many people have never taken the time to learn the basics of how an automobile works and then are shafted by unscrupulous mechanics?

I cannot change the brakes on my car. I don't know how. Again, I could buy a book, read the instructions, buy some tools, purchase a good jack, and change them, and I would then know how. I choose not to. But I know enough about the job to know if I am getting fleeced!

And if I ever decide I want to do such work, you can bet I'll learn how to do it correctly.

I can read. I can comprehend. I have limited time, just like you, but I think it is important enough that I learn all I can about the subject. I do not stand there, swaying back forth in the breeze, waiting for somebody else to do it for me. Or complain because nobody volunteers.

Maybe it is the lack of self-responsibility that seems to be so prevalent today. My generation was so conscious of protecting our kids from anything bad, making sure their precious self-esteem survived intact, and that they wanted for nothing that we raised a whole crop of, "I don't know how. Do it for me!" Or we gave them the attitude of, "I don't want to learn a skill so I can make a living for my family. I can't learn. It is too much trouble. I'm too dumb. Pay me anyway, though. It is not my fault I won't learn."

I see examples of it in the forums on the popular Ham Radio sites. Bless 'em, they do have the gumption to ask, and that is a good thing. And yes, we should all strive to be helpful.

But the post usually runs something like, "I just passed my General and spent $10,000 on a rig and amp. What kind of antenna should I buy?"

There are enough curmudgeons out there that the first replies will not be all that friendly. No, they will be downright nasty. Eventually, though, someone with a true mentor's heart will ask some questions and provide the newbie with some valid info, pointing him in the right direction to learn more, politely inviting him to search the site's archives for the hundreds of other answers to the same question, urging him to invest some time and effort in some of the myriad sources for antenna knowledge. With a little exertion, that newcomer will pick a good antenna—whether he builds it or buys it off the Internet—and learn something about antennas in general in the process. He will likely never be an RF engineer, but he will enjoy the hobby more.

Sometimes the original poster comes back with a thank-you, and a report that he or she has invested in an antenna book, visited some web sites, and is busy soldering feed line to some wire.

But too often, the follow-up post is, "What a bunch of rude SOBs! I just wanted you to tell me which antenna to buy. I don't know how to make one!"

Sigh.

As never before in the history of mankind, we are blessed with access to knowledge. I can read about any subject I can imagine…free, no waiting except for the web page to refresh…and even explore some subjects I could never have thought of before, even if I had wanted to. I recently found a site that links to a dozen different free, online Spanish courses. Want to learn about the sex life of the tsetse fly? It's there. (Just be careful how you type in that search term!) Need a manual for a piece of gear that was discontinued in 1972? Odds are you can download the PDF. Do a Google search for "antennas" and stand back!

If you want to learn about something, you can. You just have to invest the time and energy.

Let me make it clear. It is okay to ask for help. That is one of the best and most lasting traditions of our hobby. Many of us take great pride in being able to help newcomers get a good start in Amateur Radio, just as our mentors patiently helped us, back when dinosaurs roamed the earth and Marconi and I were in the same DX pile-ups.

At the same time, the mentor often learns while teaching, too. It is absolutely true that the best way to learn is to teach.

It is a cliché, but like most clichés, it is one because it is so true. Give a man a fish and you feed him for a day. Teach that man to fish and you have fed him for the rest of his life. Of course, the caveat is that the guy has to want to learn to fish.

I have infinite patience with him if he does.

I have less tolerance if he wants me to catch, clean, cook and spoon feed him that nice sea bass.

The saddest words I know: "I don't know how. Do it for me."

But maybe the most hopeful: "I don't know how. Would you help me learn?"

THE Amateur Radio dictionary

Any pursuit that mankind might adopt quickly develops its own language, its proprietary jargon, its own semi-secret dialect fully understood only by its serious practitioners.

Consider golf. Would a newcomer to the game know what a mulligan was? Or a divot? Would a novice fisherman be able to define a jig, a spinner bait, or a 30-pound-test line?

Amateur Radio is certainly no exception. The hobby has now been around for more than a hundred years and has created through the years its own dialect to be sure. Add to that the whole technical side and it becomes even more a challenge. All of this mumbo jumbo is not designed to flummox the newcomer at all. It is not some top-secret handshake to prevent others from entering a top-secret fraternity.

It is simply the way terminology develops and morphs over the years. Furthermore, with a pastime as dynamic and ever-changing as Ham Radio, it is inevitable that its verbiage changes and increases tremendously. Changes and increases so that even those who are active in the hobby may still need to be able to look up new terms as they come into common use.

Example: ask even the longest-tenured Amateur Radio operator to define "JT-65." Or to tell us what an "IOTA" is.

That, then, is the reason for this dictionary.

When I returned to the hobby after quite a few years of inactivity, I was amazed at how much the hobby had remained the same. But I was also amazed at how much it had changed. New technology, new operating events, new radiosport awards, new capabilities in our radios all made me wonder if I would ever catch up.

Many times during a QSO (see "QSO" in this dictionary) I wished for a quick-lookup that would tell me basically what it was that the guy on the air was talking about when he asked me a question or made a casual comment. Unfortunately, I often had to admit my ignorance and ask for clarification. Usually the fellow on the other end of the conversation was happy to explain it, and that gave me the opportunity to learn.

Still, I wished for some kind of Amateur Radio dictionary that would at least give me a basic idea of what the term meant and lead me to ask the right questions when someone dropped an unfamiliar word on me. As usual, I searched online for a glossary. Quite a few exist. Most are basic and woefully incomplete. Others have not been updated since the transistor. Some are totally technical and the definitions are not what I would expect to be helpful to someone new.

I can only imagine how frustrating it must be for the true newcomer to our wonderful hobby. Or for others like me who might be returning to the hobby after a hiatus. Or for that guy in a chat on 6-meter SSB who is asked what he thinks of a "hex beam" or if he is a SMIRK member.

What this dictionary is

What I have attempted to create here is a simple but more complete dictionary containing the most commonly encountered terms and jargon any of us might run across in Ham Radio. I see it as something that can be kept near the operating position in one's radio shack for quick access. Or have it downloaded on an e-book reader or tablet or computer for easy accessibility.

Language evolves. Even the brand-name English dictionaries are revised regularly to add words and terms that emerge naturally or to cull those that are no longer commonly in use. Therefore I also see this work to be a continuing thing. That is where you can help.

If you would like to suggest a new word or term to add, or question my definition, please email me at: **don@donkeith.com**. I will be happy to consider it for inclusion in this tome the next time it is updated.

What this dictionary is not

I will include a limited number of electronic or technological terms, those that the typical operator is most likely to encounter on the air or in group discussions with other Hams. Others are simply beyond the scope of this work and their inclusion might juts overwhelm the user. Besides, there are many other publications and web sites that do a fine job of that sort of thing, too.

First, I highly recommend the purchase of the American Radio Relay League's (see "ARRL") *The ARRL's Handbook for Radio Communications*. The index can lead you to easily understood descriptions and information about just about any technical term you will encounter in the hobby.

To purchase a new copy, or to find out more about Amateur Radio, visit the League's Web site at: **www.arrl.org**. You can usually find older copies of the handbook for sale at hamfests (see "hamfest"), in a boneyard (see "boneyard") or in a flea market (see "fleamarket"). They will only lack the very latest terms. Volts and ohms are still the same as they were a hundred years ago.

There are also many on-line sources for looking up electronics terms. They include, in no particular order:

http://www.maximintegrated.com/en/glossary/definitions.mvp/terms/all
http://www.csgnetwork.com/glossary.html
http://www.wilsonselectronics.net/dictionary.htm
http://www.hobbyprojects.com/dictionary/a.html
http://www.extron.com/technology/glossary.aspx

Again, this dictionary is designed as a quick lookup resource for newcomers (and old-timers, too!) to use when encountering a new or unfamiliar term while operating, listening, or reading. You may even want to simply read through at your leisure. Some term you thought you understood may mean something different altogether!

Now, if you are new to Ham Radio, welcome to the world's greatest hobby. Enjoy! Learn!

If you have been licensed since the '20s, I hope you are still enjoying the hobby as much as you did the day you first fired up that spark gap transmitter (see "spark gap") and worked your first DX (see "DX"). But even you may find some new terms here that you are not sure about.

So 73 (see "73") and BCNU (see "BCNU") on the air! SK (see "SK").

Notes:

CW abbrev.: An abbreviation typically employed by operators using CW (Morse code), though you will often encounter some of them on other modes, including voice. CW abbreviations are quite often used on digital modes, making them even more important to know.

slang: Jargon you may hear while listening or communicating on the Amateur Radio bands. While many of these terms are in common usage and not truly slang in the hobby, I will still label them as slang if they mean something different in Amateur Radio than they do in the outside world.

In some cases, and to help you better understand the usage, I will give you origins of the word or term and examples of their typical use. I will also indicate those terms that are not totally accepted on the air, and even my opinion on whether that attitude is correct or not. In most cases, I would suggest you avoid them and either use plain English or my offered alternative.

antiquated term: a word, phrase, or term that is no longer in regular use but that you may encounter on the air at some point.

Please note that I mean nothing derogatory by use of the words "slang" or "antiquated." I have to label them as something!

Also note that each section begins with the NATO phonetic alphabet pronunciation (see "NATO phonetic alphabet") and the Morse code dots and dashes (see "Morse code") for the respective letter. I use "dahs" and "dits" instead of dashes and dots because the best way to learn CW (see "CW") is by sound, not as individual dots and dashes. A dash is a "dah." A dot is a "di" or "dit."

Be aware that there is a number-and-punctuation section after the letter "Z."

Now, from A to /, here is THE Ham Radio Dictionary.

A

Alpha ("AL – fah")

Di – dah

absorption - The loss of strength of radio frequency energy as it travels through any medium. In radio, this typically applies to the strength of a signal as it passes through portions of the atmosphere.

ABT (*CW abbrev.*) – About, approximately

AC - Alternating current. The flow of electricity in which the current periodically reverses the direction of its charge. See: **alternating current**, **DC**, **direct current**

ACC – Abbreviation for "accessory."

access code - 1. Term typically used when referring to a tone or series of tones used to access or activate a particular function of a repeater station such as a link, an auto-patch, or other capability. Numbers, letters, or other symbols are entered using a telephone key pad or properly equipped microphone.
2. A sub-audible tone on a transmitted signal that is required to access a repeater station. See: **CTCSS**, **repeater**, **tone**

A/D – Analog-to-digital. The conversion of an analog signal or source to digital. See: **analog**, **digital**

ADDR (*CW abbrev.*) – Address, as in mailing or shipping location

ADIF – Amateur Radio Data Interchange Format. A vendor-neutral personal computer file format intended to allow Amateur Radio station logs to be created in a variety of software programs so that they can more easily be used by other software.

adjacent channel interference – Interference to a receiver from a station operating on a nearby frequency or channel.

Advanced – A former class of Amateur Radio license. While many current Hams hold Advanced class licenses and they can be renewed, the Federal Communication Commission is no longer issuing them.

aerial (*antiquated term*) – Antenna. The term is considered outdated though still sometimes used in Europe. See: **antenna**

aeronautical mobile - An Amateur Radio station operating aboard an aircraft.

AF – Audio frequency, the range of frequencies that can be typically detected by the human ear, generally 20 cycles per second to 20,000 cycles per second. See: **audio frequency**

AFC – Automatic frequency control. A circuit in an electronic component that automatically compensates for drift in frequency. See: **automatic frequency control**

AF gain – In Ham Radio, usually refers to a control on a receiver that allows the operator to increase or decrease the amount of amplification that is applied to the audio signal from the receiver's audio circuit. See: **RF gain**

AFSK - Audio frequency shift keying. Using audio tones generated by a computer and transmitted over a voice mode such as SSB for digital communications as opposed to employing frequency shift keying (FSK) to vary the carrier frequency. See: **audio frequency shift keying, frequency shift keying, FSK, PSK31, RTTY**

after-burner (*slang*) – An external power amplifier.

AGL – Above ground level. The term is typically used when referring to towers or antennas.

AGN (*cw abbrev.*) – "Again."

AGC - Automatic gain control. A circuit in a receiver that maintains a more constant signal as it fades or increases in strength. The reaction time of the circuit to signal changes can often be varied. See: **automatic gain control**

A-index – A daily measure of the Earth's geomagnetic activity and how it is affected by solar activity. Generally a lower number means better propagation on the HF frequencies. A-index readings can vary from single digits to over 100. See: **HF**, **propagation**

airwaves (*antiquated term*) – The frequency bands on which radio communication might take place.

ALC - Automatic level control. A circuit in a transmitter's output amplifier that helps prevent amplifier overload or a similar circuit in an external amplifier that feeds a voltage back to the exciter to avoid overdriving the external amplifier. See: **automatic level control**

Alinco – A major manufacturer of Amateur Radio equipment, headquartered in Japan. Visit: **http://www.alinco.com/**

alligator (*slang*) – 1. A term applied to stations that run high power and have a transmit signal that can be heard farther than the operator's receiving capability allows him to hear incoming signals. Origin: Citizens Band radio, where such a station was said to be "all mouth and no ears." The term can also be applied to repeater stations that transmit farther than they can typically detect and re-transmit signals. See: **CB radio**, **Citizens Band radio**, **repeater**
2. A repeater station's time-out timer that shuts off the repeater transmitter if a user talks too long. Example: "You talked too much and the alligator got you." See: **time-out**, **timer**

allocation – Channels or frequency bands designated for use by specific radio services by a country's communications regulatory agency.

all time new one (*slang*) – A country, operator, zone, grid square or other entity which the station has never worked and/or confirmed a contact. Abbreviated as ATNO. Example: "Worked Indonesia last night for an all time new one."

alternating current - The flow of electricity in which the current periodically reverses from a positive to a negative charge and back. See: **AC, DC, direct current**

AM - 1. - Amplitude modulation. A type of modulation in which the amplitude of the carrier wave is varied in relation to audio when transmitted and then converted back to audio when received. See: **AM, amplitude modulation, carrier, carrier wave, FM, modulation**.
2. (*CW abbrev.*) – Morning.

Amateur– Typically refers to a person who is licensed to operate an Amateur Radio station by the government of his or her country. By law, Amateurs are not permitted to receive compensation for their activities, thus the "amateur" designation. Example: "Sue has been a licensed Amateur since 2009." See: **Amateur Radio, Amateur Radio Service, Ham, Ham Radio, pecuniary interest**

Amateur Electronic Supply – A multi-location vendor who sells Amateur Radio equipment and supplies. Abbreviated AES. Visit: **http://www.aesham.com/**

Amateur Extra – Currently the highest available class of Amateur Radio license in the USA. Usually referred to as "Extra class." See: **General, Technician**

AmateurLogic TV – A weekly television program for Amateur Radio operators, streamed live on the Internet. Archives of past shows are also available for viewing. Visit: **http://www.amateurlogic.tv/**

Amateur Radio - A non-commercial radio service established by international treaty that allows licensees to own and operate transmitting facilities on assigned bands. In the USA, the service is regulated under Part 97 of the Federal Communication Commission Rules and Regulations. See: **Amateur, Amateur Radio Service, FCC, Federal Communications Commission, Ham, Ham Radio, Part 97**

Amateur Radio Data Interchange Format - A vendor-neutral personal computer-file format intended to allow Amateur Radio station logs to be created in a program that can more easily be used by a variety of other software systems. See: **ADIF**

Amateur Radio Newsline – A service that provides news coverage of happenings in Amateur Radio and of interest to Hams. The news is disseminated on the ARN web site, on several Amateur Radio internet programs, as well as over many repeater stations around the country. Abbreviated as ARN. Visit: **http://www.arnewsline.org/**

Amateur Radio Service - A non-commercial radio communication service established by international treaty and regulated by each nation's government. This allows citizens to become licensed to own and operate transmitting facilities using assigned portions of the radio spectrum. The service was specifically created for the purpose of self-training, communication and technical innovation and to maintain a pool of operators and stations capable of assisting in the case of emergencies and disasters. The service requires that these activities be carried out by duly licensed Amateurs solely with a personal aim and without pecuniary interest. In the United States, the service is regulated by the Federal Communications Commission under Part 97 of the Commission's rules and regulations. See: **Amateur, Amateur Radio, Ham, Ham Radio, Part 97, pecuniary interest**

Amateur Radio Supplies – A vendor of Amateur Radio equipment and supplies. Visit: **http://www.amateurradiosupplies.com/**

Amateur's Code – A suggested code of conduct for Amateur Radio operators, originally written in 1928. The Code reads:

The Radio Amateur is:

CONSIDERATE...He/She never knowingly operates in such a way as to lessen the pleasure of others.

LOYAL...He/She offers loyalty, encouragement and support to other Amateurs, local clubs, the IARU Radio Society in his/her country, through which Amateur Radio in his/her country is represented nationally and internationally.

PROGRESSIVE...He/She keeps his/her station up to date. It is well-built and efficient. His/Her operating practice is above reproach.

FRIENDLY...He/She operates slowly and patiently when requested; offers friendly advice and counsel to beginners; kind assistance, cooperation and consideration for the interests of others. These are the marks of the Amateur spirit.

BALANCED...Radio is a hobby, never interfering with duties owed to family, job, school or community.

PATRIOTIC...His/Her station and skills are always ready for service to country and community.

American Radio Relay League - The national association for Amateur Radio in the United States. Abbreviated as ARRL. The largest organization of radio amateurs in the world. According to the organization's web site, the League's mission is based on five pillars: Public Service, Advocacy, Education, Technology, and Membership. Visit: **www.arrl.org**

ammeter – A test device for measuring electrical current. See: **amp, ampere, current**

amp – 1. - Ampere, a basic unit of measurement of electrical current. See: **ampere**.

2. (*slang*) – Amplifier. See: **after-burner, amplifier, linear, linear amplifier**

ampere - A basic unit of measurement of electrical current, typically defined as the measure of the electron flow through a circuit per unit of time. The term is often abbreviated as "amps" or with the capital letters "A" or "I". See: **amp**

amplifier - A circuit that increases the voltage, current, or power of a signal. The term often describes a device used to increase the output power of a transmitter or an internal circuit within a receiver to escalate the strength of a detected radio signal. Example: "I am using an amplifier here to boost my one-hundred watts output to about eight-hundred watts." See: **amp, linear, linear amplifier**

amplitude modulation – A method of placing information on a radio signal by varying the strength (or amplitude) of the signal's carrier in proportion to an audio signal. That variation in amplitude is then changed back to audio by a circuit inside a receiver. See: **AM, modulation, receiver**

AMSAT – The Radio Amateur Satellite Corporation. An educational organization whose goal is to foster Amateur Radio's participation in space communication. AMSAT is responsible for designing, building and placing into orbit many Amateur Radio satellites. Visit: **www.amsat.org**

AMTOR - Amateur Teleprinting Over Radio. A digital communications mode in which error detection and correction are achieved by constant confirmation using "handshaking" or character repetition. AMTOR is still used by some Amateurs but has mostly been replaced by PSK31 and other more recent digital modes. See: **digital modes, PSK31**

analog – An electronic signal that carries information by a variation in time, spatial position, or voltage. See: **digital**

ANARC - Association of North American Radio Clubs. An organization made up of clubs primarily involved with shortwave radio listening. Web site: **www.anarc.org**

ancient modulation (*slang*) (*antiquated term*) – Derogatory term used for amplitude modulation (AM) by those who believe the mode causes unnecessary interference and uses too much bandwidth.

Anderson Power Poles – A commercially available type of power connectors used for quick 12-volt connections and disconnections. Visit: **http://www.andersonpower.com/products/singlepole-connectors.html**

angle of radiation – The angle at which a radio signal is emitted by an antenna and then refracted by the ionosphere. Lower angles of radiation generally result in transmissions that travel over greater distances. Higher angles may result in what is called near-vertical incidence skywave (NVIS) propagation. See: **critical angle, near-vertical incidence skywave, NVIS**

ANT (*CW abbrev.*) - Antenna. See: **antenna**

antenna - An electrical device or circuit designed to emit or receive electromagnetic radio waves. Antennas can take many forms.

antenna analyzer – A test instrument that enables checking various parameters of an antenna such as forward power, reflected power, standing wave ratio, impedance, and other factors. See: **forward power, impedance, reflected power, standing wave ratio, SWR**

antenna farm (*slang*) – Multiple antennas at a single station location.

antenna impedance – Resistance of a cable or antenna feed point in relation to the flow of electricity. Impedance is measured in ohms. Although an antenna's impedance fluctuates with the frequency of operation and many other factors, it should be approximately 50 ohms for most modern transceivers in order to achieve the maximum transfer of radio frequency energy from the transmitter to the antenna and into space. Most commercially manufactured coax cables for Amateur use have an impedance of 50 ohms. The feed point of a typical dipole antenna is approximately 50 ohms. See: **antenna, coax, dipole**

antenna matching – Employing electrical components or devices to attempt to match the impedance of an antenna system to the output impedance of the transmitter and/or receiver in use. Matching helps assure the maximum transfer of radio-frequency energy from transmitter to antenna and out into space. See: **matching**

antenna modeling – Using computer software to create diagrams and charts of expected parameters and performance of an antenna. See: **NEC, EZNEC**

antenna party (*slang*) - A get-together of Hams to assist a fellow Amateur in putting up antennas or erecting towers.

antenna pattern – A diagram of the areas where a signal is expected to be stronger or weaker for a given antenna. Such a pattern can be plotted on a graph to help visualize where a signal should be best and worst. Each antenna design has a different expected pattern, but that can be altered in the real world by local obstructions, topography, propagation and other factors.

antenna switch – A device for choosing the connection of the output/input of a transmitter, receiver or transceiver when more than one antenna is available.

antenna tuner (*slang*) – A commonly used term for a device used to match the impedance of an antenna system to the output impedance of a transmitter or input impedance of a receiver. This term is considered slang because an "antenna tuner" does not "tune" the antenna at all. It merely attempts to find values of its components so that the impedance of the antenna system is close enough to that which the transceiver requires so the maximum transfer of energy to the antenna and into space can occur. See: **auto-tuner, internal tuner, match, matchbox, mismatch, transmatch**

antipode – Two locations at exact opposite points on the Earth's surface. The antipode of your location is as far in any direction on the planet as it can be from you. If you drilled straight from your location through the Earth, you would emerge at the antipode of where you started.

anti-VOX – Circuitry in a transmitter/transceiver that prevents audio from the receiver's speaker or noise in the background in the shack from actuating the VOX, a voice-operated relay that turns transmitting on and off when the operator is speaking. See: **VOX**

apogee – A point in the orbit of a satellite in which it is farthest away from the Earth. See: **perigee**

appliance operator (*slang*) - Hams who have little interest in building or experimenting with radio equipment. Instead, they are more likely to operate commercially available equipment, often with only a minimal understanding of how it actually works.

APRS – 1. Automatic Packet Reporting System. A system for real time digital exchange of information of immediate value in the local area using Amateur Radio equipment interfaced with the Internet. The data may also be distributed globally for immediate access. This data might include messages, alerts, announcements, and bulletins, and can also be shown on a map display. Visit: **http://www.aprs.org/**

2. Automatic Position Reporting System. A capability of the Automatic Packet Reporting System in which GPS (Global Positioning System) information can be included in the shared data so the station can be tracked on maps on the Internet or on other properly equipped Amateur Radio devices.

AR (*CW abbrev.*) – "I have finished my message or transmission." Sent as one character: di-dah-di-dah-dit.

arc - An electrical flash between two conductors caused by the ionization of a vapor or gas.

ARC – The abbreviation for "amateur radio club." Example: BARC is Birmingham Amateur Radio Club.

Arduino - An open-source computer hardware and software company, project and user community that designs and manufactures kits for building digital devices and interactive objects. Many of these projects and programs can be applied to Amateur Radio uses. Visit: **http://www.arduino.cc**

ARES - Amateur Radio Emergency Service. A public service organization of the American Radio Relay League. Licensed Amateurs who have voluntarily registered their qualifications and equipment with their local ARES leadership, make themselves available for communications duty in the public service when disaster strikes. League membership is not necessary to participate in ARES but members do have to hold a valid Amateur Radio license. Visit: **http://www.arrl.org/ares**

ARISS – Amateur Radio on the International Space Station. Most crewmembers on the ISS are Amateurs licensed by their respective countries. Hams are able to chat with the astronauts during their down time and the ISS also schedules regular conversations via Amateur Radio with schools around the world. Visit: **http://www.ariss.org/**

armchair copy (*slang*) – Absolutely perfect copy of another station's transmissions.

ARN – Amateur Radio Newsline. See: **Amateur Radio Newsline**

array – An antenna system with more than one element.

ARRL - American Radio Relay League. See: **American Radio Relay League, League, The League**

AS – 1. (*CW abbrev.*) "Asia."
2. (*CW abbrev.*) "Wait for a moment, please." Sent as a single character: di-dah-di-di-dit.

ASCII - American National Standard Code for Information Interchange. A digital code for the transmission of teleprinter data. The ASCII 7-bit code represents 128 characters including 32 control characters.

ASR - Automatic send-receive. A radio-teletype (RTTY) terminal mode that allows message composition while simultaneously receiving text from another station. See: **radio-teletype, RTTY**

Associated Radio – A vendor of Amateur Radio equipment and supplies. Visit: **http://www.associatedradio.com/**

ATNO (*slang*) – Abbreviation for "All time new one." See: **all time new one**

ATT – Attenuator, attenuation. A device or circuit used to reduce the strength of a received signal, usually to block interference from a very strong and/or local station. See: **attenuator**

attenuator - A device or circuit used to reduce the strength of a received signal, usually to block interference from a very strong and/or local station. See: **ATT**

ATV - Amateur television. An operating mode in which an Amateur Radio station sends television signals over Ham frequencies. This term usually only applies to fast-scan television though Amateurs often employ a mode called slow-scan television. See: **fast-scan television, slow-can television**

audio frequency - The range of frequencies that can be typically detected by the human ear, generally 20 cycles per second to 20,000 cycles per second. See: **AF**

audio frequency shift keying - Using audio tones generated by a computer sound card and transmitted via a voice mode for digital communications as opposed to employing frequency shift keying (FSK) to vary the carrier frequency. See: **AFSK, FSK, frequency shift keying, PSK31, RTTY**

auroral propagation - Propagation of radio signals using highly ionized regions around the Earth's poles, often the Northern Lights (aurora borealis).

automatic frequency control - A circuit in an electronic component that automatically compensates for drift in frequency. Abbreviated as AFC.

automatic gain control - A circuit in a receiver that maintains a more constant signal as it fades or increases in strength. Abbreviated as AGC. See: **AGC**

automatic level control - A circuit in a transmitter's output amplifier that helps prevent amplifier overload or a similar circuit in an external amplifier that feeds a voltage back to the exciter to avoid overdriving the external amplifier. Abbreviated as ALC. See: **ALC**

automatic volume control - A circuit designed to keep a receiver's audio volume (loudness) at a constant level. Abbreviated as AVC. See: **AVC**

autopatch (*antiquated term*) - A device that interfaces an Amateur Radio repeater station to the telephone system. This allows a Ham using the repeater to make telephone calls over the air to any telephone. Sometimes simply referred to as a "patch," these devices have generally been replaced by cellular telephones. See: **patch**

auto-tuner – An antenna matching device that uses internal relays or other scheme to automatically search for the best match to the antenna system, as determined by a built-in computer rather than have the operator change the values of the device's components. The device does this automatically when it is engaged, usually with a button push or by sensing RF when the operator transmits. See: **antenna tuner, match, matchbox**

AVC - Automatic volume control - A circuit designed to keep a receiver's audio volume (loudness) at a constant level. See: **automatic volume control**

average power – The average power being run by a transmitter as measured on a standard power meter. The result is typically in watts. On some intermittent modes, such as single-sideband, the usual standard average-power meter is not physically capable of measuring the actual output power. This requires a peak-reading power meter. See: **peak-reading power meter**.

average power meter - The average power being run by a transmitter as measured on a standard power meter. The result is typically in watts. On some intermittent modes, such as single-sideband, the usual standard average-power meter is not capable of measuring the actual output power. This requires a peak-reading power meter. See: **peak-reading power meter**

AWG - American wire gauge. The standard for describing the diameter of wire. The wire size increases as the gauge number decreases.

Az/El - 1. The azimuth (horizontal) and the elevation (vertical) direction an antenna can be pointed.
2. A type of rotator that can change both the azimuth and elevation direction of an antenna. See: **rotator**

B

Bravo ("BRAH – Voe")

Dah – di – di – dit

B4 (*CW abbrev.*) – "Before."

backscatter – A form of propagation of radio waves in which those waves are reflected in the ionosphere back in the direction from which they originated. Signals so propagated can then be heard in areas they would normally skip over. See: **skip, skip zone**

balanced line - A feed line connected to an antenna that is made up of two conductors, each having equal but opposite voltages and with neither conductor at ground potential. See: **ladder line, open wire line, window line**

balanced modulator - A mixer circuit used in a single-sideband transmitter to combine a voice signal and the carrier signal, but causing the original carrier signal and half the audio signal to be suppressed. See: **single-sideband, SSB**

ball mount – A type of antenna mount with an adjustable swivel allowing an antenna to be mounted on a surface that is not horizontal or vertical. Usually used to mount an antenna on an automobile. See: **mobile**

balun – "Balanced/unbalanced." A transmission line transformer used to convert balanced input to unbalanced output or vice versa. A balun is typically used to couple a balanced antenna, such as a dipole, to an unbalanced feed line, such as coax cable, or to transition from a balanced line, such as open wire line, to an unbalanced line, such as coax cable. See: **balanced line, coax, dipole, ladder line, open wire line, window line**

band – A range of frequencies in the electromagnetic spectrum. Example: the 80-meter Amateur Radio band is between 3.5 megahertz and 4.0 megahertz.

band edge – The upper and lower limits of a band of frequencies on which an Amateur Radio station may operate. Operators should be careful to be sure no portion of their signals extend beyond the band edge.

band is changing, band is going out (*slang*) – Phrases used during a contact indicating that propagation conditions are changing and the other operator's signal is either fading or getting stronger. See: **fading, propagation, QSB**

bandpass – 1. The range of frequencies that might be allowed to pass through a filter or receiver circuit.
2. The range of frequencies that may be detected, heard, and/or displayed at any given time by a receiver.

bandpass filter - A circuit or component that passes signals in a defined range of frequencies while attenuating signals above and below that same defined range.

band plan – Frequencies that are reserved by gentlemen's agreement for specific types of operating, such as CW, DX, digital modes, AM, and more. Visit: **http://www.bandplans.com/** See: **window**

bandwidth – 1. The frequency occupied by a particular type of radio transmission.
2. The amount of data that a circuit is capable of transferring.

barefoot (*slang*) - transmitting with the normal output power of the transmitter/transceiver and not employing a linear amplifier to further boost the transmit power. See: **amp, amplifier, linear, linear amplifier**

base loading – Typically the practice of using a coil located at the bottom of a vertical antenna to raise the inductance and give the antenna a lower resonant frequency. See: **center loading**

base station (*slang*) - A radio station that is designed to be operated from a fixed location instead of being portable or mobile (in a vehicle). See: **fixed station, mobile, portable**

battery - A device that converts chemical energy into electrical energy, storing the energy, and then makes it available as needed. Some batteries are re-chargeable.

baud - The number of distinct symbol changes (signaling events) made per second in a digitally modulated signal. See: **baud rate**

Baudot - A five-bit digital code, invented by Émile Baudot, and commonly employed in teletype/teleprinter applications. Pronounced "baw – DOH."

baud rate - The measure of the speed of data transfer for a modem. Pronounced "bawd." See: **baud**

bazooka – A type of dipole antenna that uses coaxial cable as its elements. The shield of the cable is the radiating element and the center conductor acts as a matching transformer to give the antenna a wider bandwidth than a typical dipole. See: **double bazooka**

BCI – Broadcast interference. Interference caused to other services by a radio station in the broadcast service, such as commercial AM or FM stations. See: **broadcast band**

BCNU (*CW abbrev.*) - "Be seeing you."

beacon – 1. A station that transmits constant signals, usually for the purpose of navigation, homing, or determining propagation conditions.
2. A light or strobe atop a tall tower to visually warn aircraft of the presence of the structure.

beam - an antenna that offers a directional beam pattern and usually featuring some rejection of signals from the back and sides as well as signal gain in the forward direction. See: **hex beam, gain, quad, Yagi**

bent dipole (*slang*) – A dipole but with one or both elements bent at angles to fit onto a smaller plot of land. See: **dipole**

Benton Harbor lunch box (*slang*) - A small, portable transceiver kit manufactured by The Heathkit Company, which was located in Benton Harbor, Michigan. Models were sold that offered users single band AM-only coverage on either the 10, 6, or 2 meter Amateur Radio bands. They were quite popular with Hams because of their price and portability.

Beverage antenna – A long-wire receiving antenna used primarily for 80 and 160 meters. Can be hundreds to thousands of feet long, is usually hung near the ground, and is terminated in a resistor to Earth ground at one end. Named for an early developer of the antenna, Harold Beverage.

BFO - Beat frequency oscillator. A circuit in a receiver designed to create an internal signal to mix with an incoming external signal in order to produce an audio tone for CW or to inject a carrier for SSB reception.

bicycle mobile - An Amateur Radio station operating a portable station while riding a bicycle.

big gun (*slang*) – A station with lots of high-powered, expensive equipment and antennas. See: **little pistol, peanut whistle**

bird (*slang*) – A commonly used word meaning "satellite." See: **AMSAT**.

Bird (meter) – Refers to a brand of directional wattmeter. "I'll bring by my Bird and we'll check your transmitter power output."

birdie (slang) - Spurious signals that are usually produced inside a receiver itself.

BK (*CW abbrev.*) – 1. Back ("Back to you.")
2. Break in ("May I break into your QSO?")
3. Break (End of transmission.). Sent as one character: dah-di-di-di-dah-di-dah.

bleeder - A large-value resistor connected in parallel with the output of a high-voltage power supply circuit in an effort to "bleed off" the stored current in the supply's filter capacitors once the supply has been turned off.

bleed-over - Interference from another Amateur Radio station operating on an adjacent frequency or channel.

block diagram - A drawing or chart that uses rectangles and other shapes to represent major sections of electronic circuits.

BN (*CW abbrev.*) – "Been" ("BN on air 2 hours.")

BNC - A quick connect/disconnect type of coax connector commonly used with VHF/UHF equipment,

boat anchor (*slang*) – An expression usually applied to older Amateur Radio equipment, primarily because the big transformers and other components used in such gear made it very big and heavy compared to modern rigs.

bonding – 1. Using highly conductive strapping to get a better electrical connection between station equipment, towers, antennas and Earth ground rods.
2. Using copper strapping to better establish low-resistance electrical connection between various parts of an automobile for a more effective mobile installation.

boneyard – A flea market or area at a hamfest where used Amateur Radio equipment, parts, and other items may be bought and sold. See: **flea market, hamfest**

boom – 1. The part of a beam antenna that runs perpendicular to the elements and holds them in place.

2. An adjustable structure that holds a microphone above the operating position so it does not take up desk space and the mic can be easily moved side-to-side or up and down.

boom set - Headphones with the addition of a small microphone on a boom so that it can be positioned near the lips. See: **headphones, headset**

bootlegger (*slang*) – Anyone who uses on the air an Amateur Radio call sign—whether it belongs to anyone else or not—that has not been assigned to him or her. A bootlegger often does not even hold a valid Amateur Radio license. Such activity is illegal and could result in fines and jail time.

bounce - reflecting a radio signal off of an object or other medium, such as bouncing a signal off the ionosphere, a water tank, the moon, or the tail of a comet.

BPL - Brass Pounders League. See: **Brass Pounders League**

bps - Bits per second.

brag macro, brag tape (*slang*) – A pre-set computer macro in digital mode software containing information about a station and the sending operator that is played back during a contact on digital modes.

braid – 1. The woven outer conductor of coaxial cables.

2. A woven (flat) conductor which gives a large conductive area and is often used for station grounding or for electrically tying together as many of an auto's metal parts as possible for mobile operation.

3. The highly-conductive woven screen of tiny wires around a wire conductor.

brass-pounder (*slang*) – A Ham Radio operator who sends Morse code (CW) using an older keying device such as a straight key. Sometimes used to designate any operator who enjoys CW. See: **straight key**

Brass Pounders League – An organization managed by the American Radio Relay League for Morse code operators who handle large amounts of formal message traffic through the National Traffic System. Abbreviated as BPL. See: **BPL**, **National Traffic System**, **NTS**, **traffic**

breadboard – Wiring up a proposed electronic circuit on a printed-circuit board, perf board, cardboard, or other such medium in order to test the concept and performance of the circuit before committing to a more permanent design.

break – 1. A term used to interrupt an ongoing conversation, especially on a repeater station. Such usage is typically frowned upon. A station wishing to enter into a conversation should simply give his or her call sign. See: **breaker**

2. (*slang*) (*antiquated term*) – A term signifying the operator is ending a conversation with another station and starting a contact with another. Example: "Break with W8XYZ, and good morning, K4XXX." Common language is preferred.

3. In formal radio message handling, the term that indicates the preamble and sending instructions for the message are complete and the operator is going to now transmit the text of the message. The term is used again when the text of the message is complete and the operator is going to give the signature information. See: **BT**

breaker (*slang*) – An operator who desires to break into an ongoing conversation. Sometimes the operator will simply say, "Break," but this is not encouraged. The best way is for the operator to simply give his call sign. Example: "I'll stand by now for the breaker. Go ahead, breaker."

break-in – Employing circuitry when using Morse code (CW) to be able to receive signals between characters while transmitting. Full break-in enables an operator to listen to other signals between individual dots and dashes. Semi-break-in generally permits reception between characters. This allows the receiving station to interrupt the communication without waiting for the transmitting station to finish. This is sometimes referred to as "QSK," from the Q signal for "I can operate break-in." See: **full break-in, QSK, Q signal, semi-break-in**

brick (*slang*) – A small hand-held transceiver, so named because the original ones developed for the Amateur market were about the size of the typical construction brick. See: **HT, handie-talkie, walkie-talkie**

broadcast band – The parts of the radio-frequency spectrum assigned to commercial radio stations. In the USA, the AM broadcast band is 535 to 1705 kilohertz. The FM broadcast band is 87.9 to 107.9 megahertz. The channels between 87.9 and 91.9 megahertz are reserved for non-commercial and educational broadcasters.

broadcasting – Transmissions by radio or television that are intended for consumption by the general public at large. Broadcasting by Amateur Radio operators is not allowed except for bulletin dissemination and code practice by such stations as W1AW, the ARRL station. See: **W1AW**

BT (*CW abbrev.*) – "Break." 1. In formal radio message handling, the characters are sent to indicate that the preamble and sending instructions for the message are complete and the operator is going to now transmit the text of the message. The characters are sent again when the text of the message is complete and the operator is going to give the signature information. Sent as a single character: Dah-di-di-di-dah. See: **break**

2. Often sent during a CW conversation to indicate that the operator is moving from one thought or subject to another. Sent as a single character: Dah-di-di-di-dah.

BTR (*CW abbrev.*) – "Better." ("SIG BTR on this ANT.")

BTWN (*CW abbrev.*) – "Between."

bug (*slang*) – A semi-automatic key for sending Morse code that employs a spring lever to send a series of dots.

bumper mount – A device used to attach an antenna to the bumper of a vehicle.

bunny hunt (*slang*) – Using radio direction-finding equipment to locate a hidden transmitter. Such activities are usually conducted as a fun exercise or contest but skills learned can come in handy if an illegal, stolen, or malfunctioning, continuously keyed transmitter is detected and needs to be located. See: **fox hunting, RDF**

bureau – QSL bureau. Volunteer groups who help stations internationally to exchange QSL cards. They are typically maintained by a country's primary Amateur Radio organization, such as the American Radio Relay League in the USA. Stations keep the bureau stocked with self-addressed, stamped envelopes. When a DX operator works a number of stations, he or she will send a batch of cards to the bureau where they are sorted and sent on to the stations who have envelopes on file. For more on the ARRL's QSL bureau service for incoming cards, visit: **http://www.arrl.org/incoming-qsl-service**. See: **BURO, QSL, QSL card**

BURO (*CW abbrev.*) – QSL Bureau. See: **bureau, QSL, QSL card**

business communications - Any communication for the purpose of carrying on the regular business or commercial affairs of any party, something strictly forbidden on any of the Amateur Radio bands. See: **pecuniary interest**

busted call (*slang*) – An incorrectly logged call sign, typically in a contest.

C

Charlie

Dah – di – dah – dit

C (*CW abbrev.*) – "Yes," "Correct."

Cabrillo file – A computer-file-formatting scheme developed to assure radiosport operators would have a consistent way to electronically submit contest log data regardless of the software they or the contest sponsors used. Pronounced "cuh BREE oh." Visit: **http://wwrof.org/cabrillo/**

California kilowatt (*slang*) – A station that is running more output power than the legal limit

call book - a publication or CD ROM that lists all licensed Amateur Radio operators by call sign and gives their mailing addresses as they appear on file with the Federal Communications Commission. These have been mostly replaced by web sites and logging software that interface with the FCC files. See: **logging software**, **QRZ.com**

call district – The designated and numbered areas of the USA in which the Federal Communications issues call signs with that district's number to any new operator who lives within its boundaries. Example: Any station licensed in California would have the number "6" as part of its call letters, such as K6ABC. However, at one point in time, if he or she relocated to another district, a new call sign would have to be requested. Newly-licensed operators are still issued call signs with numbers that coincide with the FCC call districts but a station moving from one district to another no longer has to change call signs, nor do operators requesting vanity call signs have to request one with the district number in which they reside. See: **call sign**, **call letters**, **prefix**, **vanity call**

calling frequency - A frequency where, by gentlemen's agreement, stations may attempt to contact each other directly and then move to another frequency to continue the contact. Example: In the USA, 146.52 megahertz is the designated FM simplex calling frequency and no repeater station will have an input or output on that channel.

call letters, call sign – A unique sequence of letters and numbers assigned to and used to identify a licensed station transmitting a signal in the radio-frequency spectrum. In the USA, call signs are issued by the Federal Communications Commission. Amateur Radio call letters consist of a prefix of one or two letters, a call district number, and a one, two or three letter suffix following the number. See: **call district, prefix, suffix**

candy store (*slang*) - Term for a commercial Ham Radio equipment dealer.

cans (*slang*) – Headphones. See: **headphones, headset**

cap – Capacitor. See: **capacitor**

CAP - Civil Air Patrol, an organization that often calls on Amateurs to become members or assist in operations and drills.

capacitance - The measure of the amount of electrical charge that is held by a capacitor. Such a charge is measured in farads. See: **capacitor, condenser**

capacitor - An electronic component consisting of two or more conductive plates separated by an insulating material. A capacitor stores energy in an electric field. See: **capacitance, capacitor**

capacity hat - A system of wires or a solid metal disk attached to the top of a vertical antenna to counter its inductance, bring it closer to resonance, and increase its bandwidth.

capture effect – A phenomenon in FM transmission in which only the strongest signal can be heard even if others are transmitting on the same channel or frequency.

carbon microphone – A microphone design in which granules of carbon are used in the element to change audio to electrical energy.

carrier, carrier wave - A pure continuous radio signal with or without modulation. Such a wave can be modulated in various ways. See: **AM, FM, modulation**

carrier-operated relay - In its most common Amateur use, this is circuitry that senses a carrier on a repeater station's input frequency and causes the repeater to re-transmit that received signal. Abbreviated as **COR.**

CATV - Cable television. Delivery of television programming by cables that run into homes or other locations. In the early days of cable, the letters stood for community antenna television.

CATVI – Interference caused by or to cable television.

cavity - A very narrow filter that passes radio signals of a single frequency, usually used in repeater stations to protect the receiver from overload by a transmitter located on the same tower or nearby.

CB radio – The Citizens Radio Service, or Citizens Band. See: **Citizens Band, 11 meters**

CBR – Cross-band repeater. See: **cross-band repeater**

CC&Rs - Covenants, conditions, and restrictions. Refers to rules typically developed by homeowners' associations or real estate developers to maintain certain controls over what owners may do with their real estate properties. Generally, the goal of the CC&Rs is protect, preserve, and enhance property values in the community. However, they often restrict or prohibit a homeowner from erecting Ham antennas on his or her property.

center frequency – The frequency of an unmodulated FM signal.

center loading – The practice of placing a loading coil at the center of an antenna element in an attempt to achieve a lower resonant frequency for the antenna system.

CEPT agreement – Conference of Postal and Telecommunications Administrations (Europe). An understanding that allows Amateur Radio licensees from the USA to operate in most European countries without any further authorization or license requirement. This is different from reciprocal licensing agreements that exist with other countries. Visit: **http://www.arrl.org/cept** See: **reciprocal operating authority**

certification – Official technical approval by the Federal Communications Commission of electronic equipment intended to be sold in the USA. Except for power amplifiers, no Amateur Radio equipment requires FCC certification.

CFM (*CW abbrev.*) – "Confirm," "I confirm."

channel – A frequency or frequencies on which a station or repeater might operate. In Amateur Radio, only the 60 meter band has specific channels on which a station may transmit. Other channelization has developed as a result of gentlemen's agreements. Example: Channels are usually used in relation to FM repeater stations, and the frequency pair—such as "16/76"—is usually referred to as the repeater's channel.

charger – A device for restoring energy in a re-chargeable battery. See: **drop-in charger**

chassis - A frame or housing for an electrical device such as a TV, transceiver, power supply, computer or similar equipment

chassis ground - A wire conductor that terminates on the chassis of a device for electrical grounding purposes. See: **earth ground, ground**

CheapHam – A vendor of Amateur Radio equipment and supplies. Visit: **http://www.cheapham.com/**

check in (*slang*) – 1. (*verb*) To announce one's presence and availability in a net and to list any traffic he or she may have to send. A station should only check in at the proper time to do so as designated by the net control station.

2. (*noun*) An operator who has announced to the net control station his or her presence and availability in a network as well as any traffic that he or she desires to pass.

chirp – Rapid changes in the carrier frequency of a CW transmitter, resulting in a chirping sound on the signal. See: **CW**

choke - An inductor used to block alternating current (AC) in an electrical circuit, while passing direct current (DC).

circuit breaker - A protective component that opens a circuit when an excessive current flow occurs. Similar to a fuse but a circuit breaker may be reset, not replaced, after the cause of the excessive current flow is corrected. See: **fuse**

Citizens Band - A radio service in the USA—and many other countries—often referred to as "CB radio." It is designed for personal and business use for short-distance radio communications between individuals on a selection of 40 channels within the 27 megahertz or 11 meter band. In the USA, no license is required to transmit on the CB channels. 11 meters was once a Ham band but it was taken away to create the CB service. Some resentment lingers. However, CB has been a common stepping stone for serious radio hobbyists to move on to Amateur Radio. See: **11 meters, CB radio**

CK (*CW abbrev.*) – 1. "Check."

2. The word count in the body of a formal radio message, referred to as the "check."

CL (*CW abbrev.*) – 1. "Call sign."

2. "Call." Example: "TNX for CL."

3. "Clear." Indicates that the operator is closing down his or her station. See: **clear**

claimed score – The final score achieved in an on-air contest (radiosport) as claimed by an entrant based on his submitted log. The log is subject to inspection and updating in most contests. See: **contest**, **radiosport**

clarifier - A control on a transceiver that allows the operator to vary the receive frequency a few kilohertz either side of the VFO frequency without affecting the transmitter frequency. Sometimes known as receiver incremental tuning. See: **RIT**

CLDY (*CW abbrev.*) - "Cloudy"

clean sweep (*slang*) – A term used in radiosport (contesting) meaning contacts have been made by a participant in all possible geographic regions available to work in the event. Example: Contacts made with stations in all ARRL sections is a clean sweep in the ARRL Sweepstakes contest. See: **ARRL**, **contest, radiosport, section**

clear (*slang*) – "Clear," meaning an operator is finished transmitting and intends to close down his station. Example: "K7XXX is now clear."

Clegg – A former manufacturer of Ham VHF and UHF equipment.

CLG (CW abbrev.) – "Calling."

clicks - Undesired "clicks" or "thumps" generated by a CW transmitter as the key contacts are closed or opened. Clicks can cause interference to other stations operating on the band. See: **key clicks**

clipping - Distortion that occurs when an amplifier is overdriven and attempts to deliver an output voltage or current beyond its maximum capability. Such distortion can not only degrade the transmitting station's signal but can cause interference to other operators.

closed – 1) A condition when a frequency band no longer supports radio propagation. Example: "Ten meters was closed after about 8:30 last night."

2) The operator has shut down his station.

closed repeater - A repeater station with access limited only to a select group. See: **open repeater**

cloud warmer (*slang*) - an antenna system which tends to radiate most of its transmitted energy straight up. This is typically not a desired trait. See: **worm burner**

Clover – An Amateur Radio digital communications mode that allows full duplex communications. See: **digital mode, full-duplex**

CLR (*CW abbrev.*) – "Clear." See: **clear**

Club Log - A free Internet site for DX enthusiasts that gives them the capability of producing DXCC league tables, DXpedition tools, log search services and most-wanted lists of countries. Visit: **www.clublog.org** See: **DX, DXpedition, DXCC**

club station – An Amateur Radio station that is specifically licensed to a Ham club, established and provided for use by its licensed members.

cluster (*slang*) – A web site on which stations report hearing ("spotting") or making contact with other stations. This allows operators who wish to talk to those stations to go to that frequency and attempt to make the contact. See: **DX cluster, DX spotting, spotting**

CME – Coronal mass ejection. See: **coronal mass ejection**

coax, coax cable, coaxial cable – An unbalanced wire cable with a center conductor surrounded by insulation and a braided-wire shield, all enclosed inside an insulating jacket. The shield is designed to reduce outside electrical and radio frequency interference. Impedance values of 50 ohms and 72 ohms are typical for coax in Amateur Radio use. Pronounced "KOH ax," "koh AX ee uhl."

code - 1. (*slang*) Morse code. A method of sending text information as a series of dots and dashes using on-off tones, lights, or clicks that can be directly understood by a skilled listener or observer. Today, and in Amateur Radio, the International Morse code is the standard version in use. The header of each letter in this dictionary contains the Morse version of the respective letter. Morse code was named for its inventor, Samuel F.B. Morse. In Amateur Radio, the mode is also often referred to as CW. See: **CW, International Morse code**

2. Digital bits and bytes in which each set is decoded as a letter, number or useful character. Example: baudot. See: **baudot**

3. One of several languages used in computer programming that may be interpreted as a set of instructions by a microprocessor.

code practice oscillator – A device that creates an audio tone that may be switched on and off using a Morse code key. Used to send practice CW. Abbreviated CPO.

Code proficiency run – An opportunity to copy Morse code during special broadcasts from the ARRL station, W1AW. Correct copy submitted earns a code proficiency certificate for the speed copied. Visit: **http://www.arrl.org/code-proficiency-certificate** See: **ARRL, W1AW**

coil - A conductor that has been wound into a series of loops in order to increase the inductance in the circuit. See: **inductor**

color code – A system used to show the value of resistors or other electrical components. Numerical values are assigned to various colors that are painted onto the body of the components so the value may be determined. This is necessary because such components are often too small to have their values indicated by the actual numbers.

compression – Use of an electronic circuit that reduces the volume of loud sounds or amplifies quiet sounds by narrowing an audio signal's dynamic range. Compression is typically used by Hams to gain more loudness on their audio so they can be better heard or in an attempt to stand out in a crowd of callers. Using too much compression can result in distorted and harsh-sounding audio, which actually degrades intelligibility. See: **speech processor**

condenser (*antiquated term*) – Former term for a capacitor, an electronic component composed of two or more conductive plates separated by an insulating material. A capacitor stores energy in an electric field. See: **capacitance, capacitor**

conditions – 1. The state of the weather at the operator's location. Example: "The conditions here are cold and 28 degrees Fahrenheit."
2. Band propagation. Example: "Conditions are great this morning with signals strong from Europe."
3. The equipment being used by the operator, usually spoken as "operating conditions." Example: "My operating conditions are a Kenwood TS-590 and a dipole at 15 meters high." See: **CONDX**

CONDX (*CW abbrev.*) – 1. "Conditions." This can refer to weather, or band propagation. See: **conditions**
2. Station equipment, usually preceded by "working" or, on CW, WRKG)." Example: "WRKG CONDX HR TS-590 XCVR DIP ANT."

contact – A conversation on the air by two or more Amateur Radio operators.

contest – An on-air activity in which Ham Radio operators attempt to contact as many stations, counties, countries, zones, and/or grid squares as they can in a specified time period in competition with other Amateurs. Under the rules of each country, no cash prize or other award of value may be given for on-air contests. Winners typically receive certificates, plaques, or merely a listing in score results in publications or web sites. Visit: **http://www.hornucopia.com/contestcal/** See: **radiosport**

---

contest station – An Amateur Radio station built specifically for participating in radiosport on a grand competitive scale. Some stations are especially elaborate with multiple towers and antennas as well as operating positions for many people to man transceivers simultaneously. See: **radiosport**

Contest University – A series of seminars, webinars, and educational events for Amateur Radio operators interested in contesting/radiosport. Visit: **www.contestuniversity.com** See: **contest, radiosport**

control – A device that allows the user to vary the value or response of a component or circuit.

controller – The system or circuitry that controls a repeater station. This includes turning the repeater on and off by remote command, timing transmissions and turning off the transmitter if users talk too long, sending the station's call letters, controlling the auto patch, and programming and running the CTCSS encoder and decoder. See: **auto patch, CTCSS, repeater**

control link - The circuit used by a control operator to monitor and make adjustments to a station being operated under remote control.

control operator - An Amateur Radio operator designated by the licensee of a station to be responsible for the transmissions that are made by the station. Example: a club station. The station must be operated within the privileges granted by license to the control operator.

control point – Any location where a control operator oversees a station's operation including a station being operated by remote control.

Coordinated Universal Time - The current time at 0-degrees longitude, which passes through Greenwich, England. This time is generally used by Amateur Radio operators when logging contacts in order to avoid confusion brought on by differences in time zones around the world. Abbreviated as UTC or GMT. See: **Greenwich Mean Time, Zulu time**

cop (*slang*) – Operators who take it upon themselves to fuss over the air at other stations who do not, in the opinion of the cops, follow proper operating procedures in a DX pile-up. Cops often cause more interference than the stations they chastise. See: **pileup**

copy (*slang*) – 1. To be able to hear another Ham Radio station's transmissions. Example: "I copy you just fine."
2. To receive a formal piece of message traffic.
3. To hear and understand Morse code.

copying (*slang*) "Listening." Example: "I was copying you and Joe last night on 40 meters."

COR – Carrier operated relay. See: **carrier operated relay**

corona ball - A round ball at the top of any antenna that would otherwise have a sharp point. The object is to minimize static discharge, which could damage the antenna or any equipment attached to it.

coronal mass ejection - A massive burst of gas and magnetic energy on the surface of the sun released into the solar wind. CMEs can affect radio propagation on Earth, depending on its strength and the angle at which it arrives at the Earth's surface. Abbreviated as CME.

counter – A test instrument used to measure digitally the frequency of a tone or signal.

counterpoise – Wires, metal elements or plates, or even an automobile body that forms all or a portion of the other half of a quarter-wavelength antenna. This helps make the antenna, in effect, one-half wavelength long and closer to resonance on the frequency for which it is cut. Wire counterpoises are often called "radials." See: **quarter-wave antenna, radials**

county hunter (*slang*) – A Ham Radio operator who attempts to establish contact with stations operating in as many counties and parishes in the USA as they can. Mobile stations often become rovers to allow county hunters to make contact and nets are established to enable county-hunting enthusiasts to pursue their goal. See: **net, rover**

courtesy beep, courtesy tone (*slang*) - An obvious audible sound on a repeater station that indicates that a station using the repeater has ended his or her transmission. This is an indication to other stations that they may now transmit. The courtesy beep also usually indicates that the talk out timer has been reset. Some operators add a courtesy tone to their own transmitter so other stations will know when they have stopped transmitting but this is not encouraged. See: **talk out, talk out timer**

coverage - The geographic area in which a repeater station provides relatively reliable communications. This can vary, of course, depending on the power, antenna, topography of the area, and elevation of the station attempting to use the repeater.

CPI, CPY (*CW abbrev.*) – "Copy." Example: "CPI U 599."

CPO - Code practice oscillator. See: **code practice oscillator**

CPS (*antiquated term*) - Cycles per second. This was once the way of describing frequency of any recurring event such as a radio signal. This terminology was replaced in 1960 by the term hertz. Example: "The transmitter was operating on a frequency of 7200 cycles per second." See: **cycles per second, hertz**

CQ (*CW abbrev.*) (*slang*) – A general call made by an operator inviting anyone to answer. The term may be used on all modes, including phone, CW and digital. A CQ may also be restrictive or directional. Example: "CQ Utah, CQ Utah. Looking for any station in the state of Utah for Worked All States." Or, "CQ DX. Looking for DX stations only."

CQ Magazine – A monthly magazine for Amateur Radio operators and other radio enthusiasts. Visit: **http://www.cq-amateur-radio.com/**

CQ zones – Geographical divisions of the world determined by *CQ Magazine* for the purpose of the operating awards and radiosport events sponsored and administered by the publication. For more, visit: **http://www.cq-amateur-radio.com/** See: *CQ Magazine,* **WAZ**

critical angle - The angle at which a radio signal is refracted by the ionosphere. Lower angles generally result in transmissions that travel over greater distances. See: **angle of radiation**

critical frequency - The highest frequency at which a radio wave will return from the ionosphere rather than passing right on through into outer space. See: **maximum usable frequency**

crimp connector – A type of wire connector that relies on applied pressure to establish electrical and mechanically reliable connection rather than or in addition to soldering. A special crimp tool is typically used. See: **solder**

cross-band - Transmitting on one band while receiving on another.

cross-band repeat – Technology that allows an operator to extend the range of a low-power handheld radio by using his higher-power base station as his own personal repeater. The process requires both a dual-band mobile and base radio to retransmit on one frequency band a signal received on another frequency band, and vice versa.

cross-band repeater – A repeater station which has its input (receive) and output (transmit) on two different bands. Abbreviated as **CBR**.

CRT – A cathode-ray tube display, such as some computer monitors or television sets. This is rapidly becoming an antiquated term as CRTs are replaced by flat-panel monitors.

crystal - A piezoelectric device designed to resonate at a particular frequency. Most crystal frequencies or frequency ranges depend on the material from which the device is made, its dimensions, and the temperature at which it operates. Temperature extremes can cause a crystal to resonate at an undesired frequency, different from the one for which it was designed. See: **crystal filter, crystal oscillator, quartz crystal**

crystal filter - A network of crystals used to obtain high rejection of unwanted signals outside the range of the filter's designed operating frequency range. See: **crystal, crystal oscillator**

crystal oscillator – An electrical circuit that employs a quartz crystal to maintain as accurate and constant as possible the frequency of a transmitter or receiver. See: **crystal, crystal filter**

crystal set – A very basic radio receiver made with simple parts and a wire antenna. It requires no external power source to work, only the radio frequency energy generated by the transmitting station. Visit: **https://www.midnightscience.com/**

CSCE - Certificate of Successful Completion of Examination. This is a document certifying that a person has successfully passed one or more elements of the Amateur Radio license examinations in the USA.

CTCSS - Abbreviation for the term "continuous tone-controlled squelch system." This system uses a series of sub-audible tones on a transmitted signal to restrict access to some repeater systems. Unless the proper tone is present on the signal, the repeater will not accept the transmission. Such codes are typically used to keep distant signals from causing the repeater to re-transmit or to overcome noise or adjacent-channel interference. The tones are also employed to keep a repeater station closed except for a designated set of users. See: **access code, closed repeater, tone**

CU (*CW abbrev.*) "See you."

CubeSat – A small, cube-shaped Amateur Radio satellite designed for launch into Earth orbit. See: **AMSAT**

cubical quad – A wire antenna formed as a loop with four sides of equal dimensions. Quads usually are made up of two or more elements. Because of size, they are typically used only on the 20-meter band and up. See: **quad**

CUD (*CW abbrev.*) – "Could."

CUL (*CW abbrev.*) - "See you later."

current - the flow of electrons in an electrical circuit. See: **AC, alternating current, DC, direct current**

cut numbers (*slang*) - A way of sending numbers in Morse code (CW) by substituting shorter letter characters for the longer number characters. Typically used in contests to speed up the contact rate. Example: Instead of a report of "599," the operator would send "5NN." The number Ø is often sent simply as "dah."

cutoff frequency - The frequency at which a filter will begin to reject signals that fall outside its designed operating range.

CUZ (*CW abbrev.*) – "Because."

CW – 1. Continuous wave. This is an unmodulated, uninterrupted radio-frequency wave.
2. (*slang*) Morse code emissions or messages, even though they are an interrupted wave, broken to form CW characters. See: **code, International Morse code, Morse code**

CW Skimmer – A software program designed to de-code Morse code transmissions that can be heard within the passband of a receiver. With some software-defined radios, this can be an entire band or more worth of signals. Such information is now being voluntarily submitted by operators and made available for general viewing on a web site. Visit: **http://www.dxatlas.com/CwSkimmer/**

cycles per second (*antiquated term*) – the number of complete cycles completed each second in an alternating current or radio-frequency signal. This term was replaced in 1960 by "hertz." Abbreviated as CPS. See: **CPS**, **hertz**

D

Delta

Dah – di – dit

data communications – The transfer of data between two or more locations.

Dayton – Usually refers to the world's largest Amateur Radio get-together, the annual Dayton Hamvention in Dayton, Ohio. Visit: **http://hamvention.org/**

dB – Abbreviation for decibel. See: **decibel**

dBd – Comparison in decibels of the gain of an antenna to a theoretically ideal dipole antenna, one in a vacuum in free space with no interaction with the earth beneath it or any other potential obstructions.

dBi – Comparison in decibels of the gain of an antenna to a theoretically ideal isotropic (uniform in all directions) antenna, a point source of electromagnetic or sound waves which radiates the same intensity of radiation in all directions.

DC – Direct current. The flow of electricity in which the current flows in only one direction. See: **AC, alternating current, direct current**

DE (*CW abbrev.*) – "From," or "This is." Example: "WA4GIY DE N4KC."

deceptive signal – Any transmission by an operator, whether properly licensed or not, that is intended to mislead or confuse. Example: Someone reporting an emergency when none exists. Such transmissions are illegal in the Amateur Radio Service. See: **false signal**

decibel - A unit used to express the ratio between two values of a physical quantity one of which is typically a reference value. Often used to indicate the strength of sound or RF waves. Abbreviated as db. Pronounced "DESS uh bul."

delta loop - A variation of the loop antenna in which its continuous wire "legs" form a triangle shape. See: **loop**

demodulation - To extract the original information-bearing signal from a modulated carrier wave or signal. See: **modulation**

desense – A reduction in receiver sensitivity due to overload from a strong, nearby transmitter.

detector - The stage in a receiver in which the modulation information is recovered from a radio-frequency carrier wave.

detection – The process of converting in a receiver circuit a radio-frequency signal into a form that can be further processed by the receiving device, such as from RF to audio.

deviation – During the modulation of FM signals, the maximum amount that the frequency changes on either side of the original carrier frequency. Deviation that is too low can result in the audio being difficult to hear. Too high deviation can cause clipping of the transmitting station's audio or interference to adjacent channels.

dielectric - A non-conductive material used to separate two conductors such as the foam or plastic between the center conductor and shield in coaxial cable. Air can sometimes be used as a dielectric in such components as capacitors.

DIFF (*CW abbrev.*) – "Difference."

DigiPan – Digital Panoramic Tuning. A freeware software program that offers the capability for a panoramic display of PSK31 and PSK63 signals over a broad frequency spectrum on a computer screen. Visit: **http://www.digipan.net/** See: **PSK31**

digi-peater - A store-and-forward digital repeater station that receives and transmits a data packet on the same frequency. See: **node**, **packet radio**

digital - Devices that typically employ calculations done directly with digits instead of measurable physical quantities, such as with analog devices or communication modes. Digital signals operate using alternation between two levels that correspond to either a digit of 1 or zero. See: **analog**

digital modes – Communication schemes that employ digital means of transferring data. In Amateur Radio, these include PSK31, JT-65, RTTY, D-STAR, and more. See: **Clover**, **D-STAR**, **JT-65**, **Olivia**, **PSK31**, **radioteletype**, **RTTY**

digital signal processor/processing – Use of digital means while receiving a radio signal to improve the signal-to-noise ratio or to help hear a signal through interference to assure clearer and more legible communication. On transmit, digital signal processing is employed to improve the quality of the station's audio and transmitted signal. Abbreviated as **DSP.**

DIN plug - An electrical connector that was originally standardized by the Deutsches Institut für Normung, the German standards organization.

diode – A solid state semiconductor electronic component that allows current to pass through it in only one direction.

DIP (*CW abbrev.*) – "Dipole antenna." See: **dipole**

dip meter – An instrument used to determine the resonant frequency of an electronic circuit. See: **grid dip meter**

diplexer - A device that provides isolation between two transmit/receive ports such as antenna outputs on two different transceivers or two different antenna connections on the same transceiver. The typical use in Amateur Radio is to couple two transceivers to the same antenna, allowing an operator to receive on one transceiver and transmit on the other, or to use a single antenna with two or more outputs on the same radio. Example: You have a transceiver with an HF antenna output and a VHF antenna output and want to use a single antenna for both frequency ranges.

dipole - A basic antenna usually made of two equal lengths of wire or tubing ("elements"), joined by an insulator in the middle and fed by a two-conductor feedline at the center. Each conductor is attached to one or the other wire element. See: **antenna, element**

direct current - The flow of electricity in which the current only flows in one direction. See: **AC, alternating current, DC**

directional antenna – An antenna specifically designed to receive or emit a signal in a particular direction or directions. See: **omni-directional antenna**

director – 1. An element on a beam or other directional antenna that is located directly in front of the driven element. It "directs" radio-frequency energy in the desired direction. See: **driven element, reflector**
2. An elected official in the ARRL Field Organization, the top representative on the regional level. See: **ARRL, Field Organization**

direct path – A radio signal that travels directly from the transmitting antenna to the receiving antenna without reflecting or refracting off any other object, hill or the ionosphere. See: **multi-path**

discharge – 1. To use up all the energy provided by a battery.
2. To ground a capacitor as a safety measure to make sure any residual electrical charge stored in the component has been dissipated.

discriminator - The stage in an FM receiver in which the modulation information is recovered from the RF signal. See: **demodulation**

dish – A very directional antenna, usually round or parabolic in shape, often used at very high frequencies or for satellite work.

distortion - An undesired change in a waveform or signal.

distress call – A transmission that indicates a life-threatening situation exists. The call usually uses the terms "SOS" on Morse code or digital modes or "Mayday" on voice modes. Distress calls always have priority over any other traffic or activity that might be going on at the time. Issuing a false distress call is illegal. See: **Mayday**, **SOS**

DN, DWN 1. (*CW abbrev.*) – "Down."
2. Sent by a DX station to indicate that he or she is listening down in frequency for calls rather than on the operator's transmit frequency. Example: "DN 2" means the station is listening down approximately 2 kilohertz for stations to call. See: **down**, **DX**, **split**, **UP**

DominoEX – A type of digital mode that uses rapid frequency-shift keying and is nominally twice as fast as PSK31.

dongle (*slang*) – A small piece of hardware that attaches to a computer, television, Ham rig, or other electronic device in order to enable additional functions.

Doppler effect, Doppler shift – The change in frequency of a wave (or other periodic event) to an observer as the source moves relative to the position of the observer. In Amateur Radio, this is most often noted when receiving a signal from an orbiting satellite, requiring the listener to continually vary the receive frequency to keep the satellite tuned in.

double (*slang*) – When two or more stations inadvertently transmit simultaneously on the same frequency or channel.

double bazooka - A type of dipole antenna that uses coaxial cable as its two elements. The shield of the cable is the radiating element and the center conductor acts as a matching transformer to give the antenna a wider bandwidth than a typical dipole. See: **bazooka, coax**

double nickels (*slang*) - A term often used on DX or county-hunting nets or in normal contacts meaning a signal report of "55." Origin: Citizens Band radio. See: **county hunter, RST, signal report**

doublet – Another word for dipole antenna. See: **dipole**

doubling (*slang*) – Two or more stations transmitting simultaneously on the same frequency or channel.

down (*slang*) – An indication by a highly sought DX station that the operator will be listening down in frequency for calls. This keeps the large number of callers from interfering with the DX station. Example: "I'm listening down five." See: **DN, DWN, DX, split, Up**

downlink– The frequency a repeater station or satellite uses to transmit to a user. See: **uplink**

DR (*CW abbrev.*) - "Dear." "My friend." Example: "TNX QSO DR BOB."

Drake – A former manufacturer of Amateur Radio Equipment. The R. L. Drake Company.

D-region, D-layer – In the Earth's atmosphere, the lowest region of the ionosphere, located approximately 25 to 55 miles above the planet. It disappears very quickly after sunset and is slow to regenerate after sunrise, especially on short winter days. The D-layer's primary effect on radio propagation is to absorb energy from the signals as they pass through it.

drift (*slang*) – A slow unintentional and undesired change in the frequency of a transmitter or receiver.

drive – 1. (*noun*) The RF power applied by a transmitter's internal oscillator to the radio's final amplifier stage.

2. (*verb*) To use an internal oscillator to supply RF power to a transmitter's final amplifier stage.

3. (*verb*) To provide RF from a transmitter to an external amplifier to achieve higher power output. See: **driver, oscillator**

driven element – The antenna element that is typically connected directly to the feed line. This term is most often used in beam or other directional antenna systems. See: **beam, director, reflector, Yagi**

driver – 1. The stage in a transmitter that initially provides RF to the radio's internal final power amplifier section.

2. The transmitter/transceiver being used to provide RF to an external power amplifier.

3. An electrical circuit or electronic component that is employed to control another circuit or component.

drop-in charger – A device for restoring energy in a re-chargeable battery for a hand-held transceiver. The radio or its detached battery can simply be dropped into a slot in the device until the battery is fully charged. See: **charger**

dropping out (*slang*) - A repeater station is no longer able to hear a station and is unable to attempt to re-transmit the signal. Example: "Sorry, you are dropping out of the repeater and I can't copy you."

DSP – Digital signal processor or digital signal processing. See: **digital signal processor/processing**

D-STAR – Abbreviation for Digital smart technologies for Amateur Radio. D-STAR is a digital voice and data protocol specification mostly used in VHF/UHF transceivers manufactured by Icom. As of this writing, other manufacturers hint they may soon produce D-STAR gear. Visit: **www.icomamerica.com/en/products/amateur/dstar/dstar/default.aspx** See: **digital, digital modes**

DSW (*CW abbrev.*) – "Goodbye," when conversing with a Russian Ham. Abbreviation for *dos vadanya*, Russian for "Until we meet again."

DTCS – Abbreviation for digital tone-coded squelch. A selective call system in which tones are used to allow the incoming signal to overcome the restrictive squelch of a receiver. See: **squelch**

DTMF – Abbreviation for dual tone multi-frequency, commonly referred to as "touch-tone." A set of dual audio tones used to send and receive numeric information, such as telephone numbers, or for actuating remote- control commands.

dual band – An antenna or radio that is designed for use on two different Amateur Radio bands. See: **multi-band**

dummy load – A non-radiating load connected to a transmitter to enable testing without interfering with other stations.

dupe (*slang*) - A duplicate contact. The term is typically used in a contest in which the rules forbid contacting the same station more than once for score credit. Example: "Sorry but we worked before on this band. You are a dupe." Pronounced "dewp." See: **contest, radiosport**

duplex – An operating mode in which the transmit and receive frequencies are not the same, such as with repeater stations. See: **simplex**

duplexer - A device that allows a single antenna to transmit and receive simultaneously. The device provides isolation on a system on the same band. Example: a repeater receives a signal on 146.28 megahertz at the same time that it is re-transmitting that signal on 146.88 megahertz. A duplexer makes it possible to do this without the transmitter interfering with the receiver.

duty cycle - The proportion of time during which a component, device, or system is operated. Example: The duty cycle of a transceiver is much shorter when using constant-carrier modes such as FM or RTTY as opposed to intermittent modes such as SSB.

DVM - Digital voltmeter.

DWN (*CW abbrev.*) – "Down." See: **down, DN**

DX – 1. (*CW abbrev.*) – "Distance."
2. (*slang*) (*noun*) A distant station or a station outside one's own country or region.
3. (*slang*) (*verb*) To attempt to contact stations in other countries via Amateur Radio. Example: "I like to DX on 17 meters."

DXCC – DX Century Club. See: **DX Century Club**

DX Century Club - An operating award sponsored by the American Radio Relay League. Abbreviated as DXCC. To earn the award, you must contact at least 100 different countries or DX entities and then confirm those contacts through Logbook of the World or by receiving QSL cards for each. Endorsements are also available for confirming contact with more DX entities. Visit: **http://www.arrl.org/dxcc** See: **DX entities, Logbook of the World, LOTW, QSL card**

DX cluster - A web site on which stations report hearing ("spotting") or making contact with other stations. This allows operators who wish to talk to those stations to go to that frequency and attempt to make a contact. See: **cluster, DX spotting, spotting**

DX Engineering - A vendor of Amateur Radio equipment and supplies. Visit: **http://www.dxengineering.com/**

DX entity – A location designated as a "country" for the purpose of operating awards or contesting. For example, Alaska, Hawaii and Puerto Rico are all considered to be "countries" or DX entities.

DXer (*slang*) – 1. An Amateur Radio operator who actively pursues contacts with Amateur Radio stations in other countries, and especially in those countries in which there are few operators and with whom contacts are rare.

2. A hobbyist who listens for shortwave broadcasts from stations in other countries and attempts to obtain confirmations from those stations. See: **shortwave listener**

DXLab - A popular software program for logging Amateur Radio on-air contacts and for use in radiosport events. The program is offered free of charge. Visit: **http://www.dxlabsuite.com/**

DX-pedition, DXpedition – An organized Amateur Radio operation from a foreign country, usually an entity in which there are few native Hams, giving operators an opportunity to establish and confirm contact with that country.

DX spotting - A process in which stations report hearing ("spotting") or making contact with other stations to a web site. This allows operators who wish to talk to those stations to go to that frequency and attempt to make the contact. See: **cluster, DX cluster, spotting**

DX Store, The – A vendor of Amateur Radio equipment and supplies. Visit: **http://www.dxstore.com/**

DX University - A multi-media program offering information, instruction and learning opportunities for DXers. Visit: **http://www.dxuniversity.com/** See: **DX, DXer**

DX window (*slang*) - A range of frequencies set aside by gentlemen's agreement to allow for foreign Amateur Radio stations to call CQ and work other stations around the world while those in the United States and Canada refrain from other types of activity there. The DX stations may also invite U.S. and Canadian stations to call them. Example: 3.790 – 3.800 megahertz is the 75 meter DX window. Visit: **http://www.bandplans.com/**

dynamic range - The ratio between the largest and smallest possible values of a changeable quantity, such as in radio signals or sound reproduction. In Amateur Radio, this most often involves how a receiver handles very weak or very strong signals.

E

Echo

Dit

E – Symbol for electromotive force or voltage, measured in volts.

earth ground - A wire conductor that terminates in the earth for electrical purposes, typically using a ground rod. See: **chassis ground, ground, grounding**

Earth-moon-Earth - Using the moon as a passive reflector off which to bounce signals back to Earth at some distance from the originating station. Sometimes referred to as moonbounce. Abbreviated as EME.

earth station - An Amateur Radio station located on or near the Earth's surface that is used to communicate with space stations such as the International Space Station, or to talk with other stations located on Earth by means of satellites in orbit.

echo – An instance when a received signal arrives at the same location but concurrently by different paths, one taking slightly longer to travel than the other, thus creating what sounds like an echo.

Echolink – A software system that allows licensed Amateur Radio stations to communicate with one another over the Internet through properly equipped repeater stations, using streaming-audio technology and a computer or telephone device. Visit: **http://www.echolink.org/**

eHam.net – A popular web site for Amateur Radio operators. See: **www.eham.net**

EHF - Extremely High Frequency, typically 30 to 300 gigahertz. See: **ELF, HF, UHF, VHF, VLF**

EL (*CW abbrev.*) – "Element." Example: "Using 2 EL beam."

E-layer - The region of the ionosphere in the Earth's atmosphere that is located approximately 55 to 90 miles above the planet. This layer normally absorbs radio signals passing through it. During certain times of the year, however, this layer becomes ionized and will refract signals on higher frequencies, propagating them over greater distances than usual. See: **sporadic-E**

Elecraft – A major manufacturer of Amateur Radio equipment, headquartered in Aptos, California. Visit: **http://www.elecraft.com/**

Electric Radio Magazine – A magazine primarily for Amateur Radio and electronic hobbyists that concentrates on vintage equipment and radio history. Visit: **http://www.ermag.com/**

electrolytic capacitor - A type of capacitor used most often as power supply filters. Such capacitors do require attention to their polarity and are capable of storing enough of an electric charge to be dangerous, even after they have been disconnected from power. Some types of electrolytics also tend to dry out over time and may fail if they have not been used in a long time.

electronic keyer - A circuit used for creating the dots and dashes of Morse Code semi-automatically using a key that is commonly referred to as a "paddle." Dots are made by pressing the paddle one way and dashes by pressing the paddle the other. See: **keyer, paddle**

electromagnetic pulse - A high-energy magnetic field which can be caused by a number of natural occurrences such as a lightning strike or by events such as a nuclear explosion. It could cause severe damage to electronic equipment, power distribution systems, and more. See: **EMP**

electromotive force - The force or pressure that pushes an electrical current through a circuit. Voltage. Abbreviated EMF and represented mathematically by the letter "E." See: **voltage**

element – The conductive parts of an antenna. In a dipole, these are the two wires that are separated by an insulator in the middle. In a beam, these are the parallel conductors attached perpendicular to the boom. See: **beam, boom, dipole**

elephant (*slang*) - a repeater station that receives from a much greater distance than it can transmit, so named because the animal has big ears but a small mouth!

elevation – 1. When describing a beam antenna, the angle at which it is aimed in reference to the horizontal.
2. How high an antenna, tower, satellite or other object is, often referenced in communications to either sea level or height above average terrain in the nearby area. See: **HAAT**

ELF - Extremely low frequency. See: **extremely low frequency**

elmer (*slang*) – An Amateur Radio mentor, usually an experienced operator who assists newcomers to the hobby.

EmComm – Emergency communications. See: **emergency, emergency traffic**

EME - Earth-moon-Earth. See: **Earth-moon-Earth**

emergency – An instance in which there is an imminent threat to life or property.

emergency traffic - A message or communication passed along from one station to another that involves the life or safety of persons. Such traffic always has priority and all stations must standby until it has been completed. See: **informal traffic, NTS, National Traffic System, priority traffic, traffic**

EMF - Electromotive force, another word for electrical voltage. See: **electromotive force, voltage**

EMI - Electromagnetic interference, a disturbance that affects an electrical circuit due to radiation emitted from an external source.

emission – An electromagnetic signal.

emission mode, emission type – The specific type of electromagnetic signal being generated by a transmitting device. Examples: AM, FM, or single-sideband. Emission modes are more completely described and defined by regulatory agencies in each country, such as the Federal Communications Commission in the USA.

EMP - Electromagnetic pulse. See: **electromagnetic pulse**

EPROM, EEPROM – Electrically erasable programmable read-only memory. A digital chip designed for storing and accessing data.

eQSL – A website on which contacts between two Amateur Radio stations may be confirmed electronically and digital QSL "cards" exchanged. Visit: **https://www.eqsl.cc/** See: **QSL card**

ERP - Effective radiated power. The total amount of radio-frequency power being emitted at the antenna considering actual power from the final amplifier stage, feedline loss, and antenna gain.

ES (*CW abbrev.*) – "And."

E-skip – Propagation of signals using sporadic-E-layer refraction. See: **E-layer, sporadic E**

ESSB – Extended single-sideband. See: **extended single-sideband**

ether (*antiquated term*) (*slang*) – Prior to the discovery of the ionosphere and its effect on radio waves, conventional theory held that signals were propagated by a substance in the air called ether. The term is still sometimes used as jargon when speaking of signal propagation.

EU (*CW abbrev.*) – "Europe."

EVE (*CW abbrev.*) – "Evening."

exam cram (*slang*) – A one- or two-day session in which instructors go over and explain correct answers to Amateur Radio exam pool questions with prospective Hams. At the end of the session(s), Volunteer Examiners administer the licensing exams. See: **question pool**, **Volunteer Examiner**

exam question pool – See: **question pool**

exam session - An event in which the Amateur Radio license examination is administered by Volunteer Examiners. See: **exam cram**, **VE**, **Volunteer Examiner**

exchange – The information passed between two Amateur Radio stations while in contact with each other. Might include signal reports, name, location, and equipment being used plus more. A far shorter exchange is typically used in radiosport. Example: "The contest exchange consists of signal report and state." See: **radiosport**

exciter – 1. The oscillator and—if using voice modes—the modulator in a large transmitter.
2. A transmitter or transceiver used to drive an external power amplifier.

extended single-sideband - Experimentation with the audio quality of single-sideband transmissions to attempt to achieve high fidelity sound within the limitations of the mode. Abbreviated as ESSB. Visit: **http://www.nu9n.com/essb.html**

Extra class (*slang*) – The Amateur Extra class of Amateur Radio license. See: **Amateur Extra**

Extremely Low Frequency – Part of the radio-frequency spectrum, usually designated as 3 to 30 hertz. While ELF has use in weather science and the medical field, it is generally thought of in relation to communication with submarines. See: **ELF**

eyeball, eyeball QSO (*slang*) - A face-to-face meeting between two Ham Radio operators.

EZNEC – Brand name of a popular software program for modeling antenna design. Visit: **http://www.eznec.com/** See: **antenna modeling, Numerical Electromagnetics Code**

F

Foxtrot

Di – di – dah – dit

FAA - Federal Aviation Administration.   See: **Federal Aviation Administration**

fading – A reduction in the strength of a received signal due to atmospherics or other factors.

false signal – Any transmission by an operator, whether properly licensed or not, that is intended to mislead or confuse. Example: Someone reporting an emergency when none exists. Such transmissions are illegal. See: **deceptive signal**

Family Radio Service – An unlicensed communications service for short-range communications by friends or family members using hand-held radios. The service uses the UHF range so does not suffer the interference that plagues the 11-meter Citizens Band. Abbreviated as FRS.

fan dipole – An antenna that uses one feed point for several dipole antennas, each cut to operate on a specific frequency band. Each shorter antenna fans out beneath the one above it, thus the name.

Farnsworth method - a way of learning and sending Morse code characters in which each character is sent at a faster rate but spaces are left between each character to effectively lower the word-per-minute rate. Example: Characters are sent at 15 words per minute but the spacing is adjusted so that the overall code speed is 5 words per minute. This often helps those learning Morse code to increase reception speed more quickly as they practice.

fast-scan television – A mode that allows Amateur Radio operators to send and receive live-action TV images similar to analog broadcast television. See: **ATV, FSTV**

FB (*CW abbrev.*) – "Fine business," "Excellent," "Good," "Fine."

F/B – Front-to-back. See: **front-to-back ratio**

F connector – An inexpensive coax connector, used primarily on smaller size cables for television and cable TV.

FCC - Federal Communications Commission. See: **Federal Communications Commission**

FDIM – Four Days in May. See: **Four Days in May**

Federal Aviation Administration - The agency in the USA that regulates all facets of air transportation. This includes such areas as tower height restrictions and lighting requirements. Abbreviated as FAA.

Federal Communications Commission - The governmental agency that regulates Amateur Radio in the USA. Abbreviated as FCC.

Federal Registration Number - An identification number assigned to individuals by the Federal Communications Commission for use when accessing the Commission's on-line site and user accounts. Abbreviated as FRN. The number can be used while applying for, modifying or renewing an Amateur Radio license.

feeder - A transmission line used to transfer the power from a transmitter to the antenna. Usually a coaxial cable or open-wire line. Often referred to as a feedine. See: **feedline, transmission line**

feedline – The wire or cable that connects a radio to an antenna.

fender mount – A device for affixing an antenna to the fender of a vehicle.

FET - Field-effect transistor. A type of transistor that is typically used as an amplifying device.

FER (*CW abbrev.*) – "For."

Field Day – An annual Amateur Radio operating event in June in which groups set up stations in portable locations to practice emergency communications. Field Day is sponsored by the American Radio Relay League and is the hobby's largest event in terms of participants. See: **operating event**

Field Offices – Branch offices of the Federal Communications Commission, which is headquartered in Washington, DC. Regional offices are in Chicago, Kansas City, and San Francisco. There are 16 district offices and 8 resident agent offices around the country. Personnel in these offices conduct on-scene investigations, inspections and audits, respond to safety of life issues, and investigate and resolve interference complaints and violations in all communications services. See: **Federal Communications Commission**

Field Organization – Volunteers who work with the American Radio Relay League to assist other Hams. Some are elected by ARRL members while others are appointed and represent constituents at the regional, state/section and local levels. See: **ARRL**

field strength meter - A test instrument used to show the presence of and measure the strength of radio frequency energy in the area where the meter is located.

filter – A circuit or device designed to only allow certain specific frequencies to pass through.

final – 1. The last transmission by a station during a contact. Example: "This will be my final. Good night!"

2. The last amplifying stage of a radio transmitter or external amplifier.

3. The amplification tube or transistor that is a part of a transmitter's final output stage. Example: "The finals in my amp are a pair of 811As."

first personal (*slang*) – An operator's first name. This term originated as Citizens Band jargon but is usually frowned upon by Hams. Most prefer simply "name" or "handle."

fist (*slang*) – The personal and individual characteristics of how an operator sends Morse code characters. An operator with a good "fist" is one whose sending is easily de-coded.

FISTS - An organization of CW enthusiasts whose mission is to promote the use of Morse code on the Amateur Radio bands, encourage newcomers to learn and use CW, and to sponsor activities to increase interest in the mode. Visit: **http://www.fistsna.org** See: **CW**

five-by-five (*slang*) – A signal report giving a signal strength of 5 and a readability of 5 on the RS scale. Sometimes referred to as "double nickels." See: **double nickels, RST**

fixed station - A radio station that is designed to be operated from a fixed location instead of being portable or mobile (in a vehicle). Sometimes called a "base" or "base station." See: **base station, mobile, portable**

flat-top antenna (*antiquated term*) (*slang*) – A dipole antenna hung so the ends and middle are at the same height above the ground. See: **dipole**

flat topping – Over-modulating a carrier signal so badly that it distorts the waveform when viewed on an oscilloscope.

F-layer- The region of the ionosphere between about 90 and 400 miles above Earth. This area of the Earth's atmosphere is responsible for most long distance radio signal propagation on the frequency bands below 30 megahertz. During hours when the Earth is in sunlight, solar heating can cause the F-layer to split into two separate layers, F1 and F2.

FLdigi – Software digital modem for receiving and sending Amateur Radio digital modes. Visit: **http://www.w1hkj.com/** See: **digital modes**

flea market – An area at a hamfest or other gathering where used Amateur Radio equipment, parts, and other items may be bought and sold. See: **boneyard, hamfest**

Flexradio Systems – A major manufacturer of software defined radios for Amateur Radio, headquartered in Austin, Texas. Visit: **http://www.flexradio.com/**

flutter - Rapid variation in the strength of a received signal, normally caused by a variation in the propagation of the signal.

FM – 1. Frequency modulation. See: **frequency modulation**
2. (*CW abbrev.*) – "From."

FMRE - Mexican Federation of Radioexperimentadores. See: Mexican Federation of Radioexperimentadores

FOC - First Class CW Operators Club. A worldwide group of Amateur Radio operators that attempts to promote good CW (Morse code) operating habits, activity, friendship and socializing on the bands. Visit: **http://www.g4foc.org**

folded dipole – A dipole antenna in which the ends of the wires have been folded back and run parallel to the original wires, forming a very "skinny" loop. Television twin-lead or ladder line is often used in construction. The antenna typically offers wider bandwidth than a traditional dipole. See: **dipole, twin-lead**

formal traffic (*slang*)– A message or communication that follows established format and is passed along from one station to another in a pre-prescribed manner. Such traffic may also carry a "precedent," such as "routine," "priority," or "emergency. See: **emergency traffic, informal traffic, NTS, National Traffic System, precedence, priority traffic, traffic**

forward power – The power radiated by an antenna less any reflected power. See: **antenna analyzer, reflected power, SWR, standing wave ratio**

Four Days in May – An annual seminar specializing in QRP interests, sponsored by the QRP Amateur Radio Club International. It is held in Dayton, Ohio, and coincides with the Dayton Hamvention. Abbreviated as FDIM. Visit: **http://www.qrparci.org/fdim** See: **Dayton, QRP ARCI**

fox hunt – Using radio direction-finding equipment to find a hidden transmitter. Such activities are usually conducted as a fun exercise or contest but skills learned can come in handy if an illegal, stolen, or hung-up transmitter is detected and needs to be located. See: **bunny hunt, RDF**

FREQ (*CW abbrev.*) – "Frequency."

frequency – Generally, the rate of oscillation of a wave. The frequency of audio and radio waves is measured in hertz (cycles per second). See: **hertz, oscillator**

frequency coordinator – A group or an individual responsible for coordinating and assigning frequencies for repeater stations, assuring minimum interference to existing repeaters.

frequency modulation - A mode in which the encoding of information on a carrier wave is accomplished by varying the instantaneous frequency of the wave. Abbreviated as FM.

frequency shift keying - A methodology in which digital information is transmitted through discrete frequency changes of a carrier wave. Abbreviated as FSK. In Amateur Radio, FSK has primarily been used with such digital modes as PSK31 or RTTY but is being replaced more now with audio frequency shift keying or AFSK. See: **AFSK, audio frequency shift keying, FSK, PSK31, radioteletype, RTTY**

Friendly Candy Company (*slang*) – Term sometimes used for the FCC, the Federal Communications Commission. See: **FCC, Federal Communications Commission**

FRN – Federal Registration Number. See: **Federal Registration Number**

front end – The circuitry in a receiver that first encounters a signal at the input, usually including all components that act on an incoming signal before its frequency is altered for more processing.

front-end overload – When a very strong signal overpowers a receiver's early stages, causing interference.

front end protector – A circuit which can guard against a receiver being overloaded or damaged by very strong signals appearing at its input. Of special value in a situation when other Amateur Radio stations are operating very close by.

front-to-back ratio - The ratio of an antenna's gain in the forward direction to that in the opposite direction. Most often a factor in beam antennas and directional arrays.

front-to-side ratio - The ratio of an antenna's gain in the forward direction to the gain at right angles—off the sides—to the forward direction. Most often a factor in beam antennas and directional arrays. Abbreviated as F/S.

FRS – Family Radio Service. See: **Family Radio Service**

F/S – Front-to-side. See: **front-to-side ratio**

FSK - Frequency shift keying. See: **frequency shift keying**

FSTV - Fast scan TV. See: **ATV, fast scan TV**

full break-in – Employing circuitry when using Morse code (CW) to be able to receive signals, even between dots and dashes, while transmitting. This allows the receiving station to interrupt the communication without waiting for the transmitting station to finish sending. This is sometimes referred to as "QSK," from the Q signal for "I can operate break-in." See: **break-in, QSK, Q signal, semi-break-in**

full duplex – A mode of operation in which communication takes place on two different frequencies simultaneously. An example is a typical telephone conversation.

full gallon (*slang*) – Running the maximum amount of legal power from an amplifier.

full quieting – A condition on FM transmissions in which an incoming signal is sufficient to completely cover any other noise in the receiver.

FUNcube – A series of educational Amateur Radio satellites built and maintained by Hams in Great Britain, the Netherlands and others under the banner of AMSAT-NA and AMSAT-UK. Visit: **http://funcube.org.uk/** See: **AMSAT, CubeSat**

fundamental - The lowest frequency or band of frequencies to which a harmonic frequency or band is related. See: **harmonic**

fuse – A safeguard component designed to fail and open the circuit if the current being drawn in an electrical circuit exceeds the maximum rated current for the device. The fuse must be replaced with a new one once the cause of the excessive current is corrected. See: **circuit breaker**

G

Golf

Dah – dah – dit

G (*slang*) – An Amateur Radio operator from Great Britain. Many British call signs begin with the letter "G." Example: "I had a nice chat on 40 meters with a 'G' from Liverpool."

G5RV – A form of dipole antenna that uses a specific length of balanced feedline between the feedpoint and the rest of the coax feedline as a matching stub in order to achieve a better match on several bands. Originally developed by British Amateur Louis Varney G5RV. See: **balanced line, match, matching stub**

GA - 1. (*CW abbrev.*) – "Go ahead."
2. (*CW abbrev.*) "Good afternoon."

gain – 1. In antennas, the increase in the effective power radiated by or in received signal strength from an antenna in a certain desired direction or angle. See: **beam**
2. Increase in received or transmitted signal strength from an amplifier.

gallon (*slang*) – Transmitter output power. Example: "I'm running a half gallon here." (750 watts, or half the maximum allowed, 1500 watts PEP, a full gallon.) See: **full gallon**

GB (*CW abbrev.*) – "Goodbye."

GD (*CW abbrev.*) – 1. "Good day."

2. (*CW abbrev.*) "Good."

GE (*CW abbrev.*) – "Good evening."

gel cell - A sealed lead-acid rechargeable battery which uses a chemical gel to generate an electrical charge.

General class – The current intermediate class of Amateur Radio license available in the USA. See: **Amateur Extra, Technician**

general-coverage receiver - A receiver that is capable of hearing a very wide range of frequencies. Typically, a general-coverage receiver will tune continuously from the AM broadcast band (535 kilohertz) to 30 megahertz, including the Ham bands. See: **Ham-band-only receiver**

General Mobile Radio Service - A licensed North American FM UHF radio service for short-distance two-way communication for family use, typically within a city. Abbreviated as GMRS.

generator - A device that converts mechanical energy to electrical energy for use in an external circuit. In Amateur Radio, this is typically a gasoline-powered internal combustion engine that produces 110 volts AC current for use in situations when normal commercial power is unavailable.

GESS (*CW abbrev.*) – "Guess."

Gigaparts – A vendor of Amateur Radio equipment and supplies. Visit: **http://www.gigaparts.com/**

gin pole – A device attached temporarily to a tower to enable raising and lowering tower sections while constructing the tower or antennas or other objects while attaching or detaching them from the structure. It usually includes a rope or wire long enough to reach the ground below, a winch or pulley, and a strong clamp for attaching the device to the tower while in use.

GG (*CW abbrev.*) – "Going."

GL (*CW abbrev.*) – "Good luck."

GM (*CW abbrev.*) – "Good morning."

GMRS – General Mobile Radio Service. See: **General Mobile Radio Service**

GMT – Greenwich Mean Time. See: **Greenwich Mean Time, Coordinated Universal Time, Zulu time**

GN (*CW abbrev.*) – "Good night."

GND (*CW abbrev.*) – "Ground."

go ahead (*slang*) – "I am completing my transmission. Go ahead with yours."

go-kit (*slang*) - A complete self-contained portable station including transceiver, power supply and/or battery, antenna and personal supplies that allows a Ham to quickly go and set up for a portable operation or to report where needed in an emergency situation.

Gooney Bird (*slang*) (*antiquated term*) – A small, low-power two-meter AM transceiver once manufactured by the Gonset Company.

GOTA station – "Get on the Air" station. A station set up for ARRL Field Day to be operated by unlicensed visitors and newcomers to the hobby under the supervision of a control operator. See: **Field Day**

grace period (*slang*) - The time period allowed by the Federal Communications Commission after an Amateur Radio license has expired in which the licensee may still renew that license. No on-air operation is allowed during the grace period. If the license is not renewed by the end of the period, the examination must be retaken to once again become licensed.

gray line – The transition region between day and night or night and day that constantly moves around the globe. Signals are often strongest when both ends of a conversation are in this day/night or night/day zone.

great circle route - The shortest path between any two points on the planet Earth. This would be the direction an operator would typically want to aim a beam antenna to send and receive the strongest signal to that area. See: **long path**

green stamp (*slang*) – One dollar bill in USA currency. Generally one or two dollars along with a self-addressed envelope are sent along with a QSL card request to an operator in a foreign country in order to offset postage for a return card. (Note: the self-addressed envelope accompanying the request is not stamped since USA-issued stamps are not honored in most countries.) See: **IRC**

Greenwich Mean Time - Coordinated Universal Time. The time at 0-degrees longitude, which passes through Greenwich, England. This time is generally used by Amateur Radio operators when logging contacts in order to avoid confusion brought on by differences in time zones around the world. See: **UTC, GMT, Greenwich Mean Time, Zulu time**

grid dip meter - An instrument used to determine the resonant frequency of an electronic circuit. See: **dip meter**

grid square - An alphanumeric geographical coordinate system (usually given as four or six characters), based on the Maidenhead Locator System developed by VHF/UHF enthusiasts. The entire globe is divided into rectangles denoted by alphanumeric codes. These grid squares are often used in VHF and UHF DX activities and many radiosport events. Visit: **http://www.arrl.org/grid-squares**

ground - 1. An electronically neutral circuit that has the same electrical potential as the earth that surrounds it. See: **earth ground**
2. A non-current carrying circuit designed to provide electrical safety.

3. A point of reference within an electrical circuit or system. See: **chassis ground**

4. The negative side of an electrical circuit.

5. The negative side of a battery.

grounding – electrically connecting equipment or antennas to an earth ground. See: **earth ground**

ground-plane antenna - a vertical antenna typically built with the vertical radiating element one-quarter-wavelength long and with several radials slightly longer than one-quarter wave extending horizontally from the base.

ground rod - A highly-conductive rod that is driven into the earth to create a ground for electrical equipment. For Amateur Radio stations, a heavy copper wire or strap is run from the station equipment—which is bonded together with strap—to the ground rod. See: **bond**

ground wave - radio waves that tend to travel along the surface of the earth, often beyond the horizon.

GUD (*CW abbrev.*) – "Good."

guy – A set of ropes or wires used to secure a tower or mast once it has been erected. Sometimes mispronounced as "guide wires."

GV (*CW abbrev.*) – "Give."

H

Hotel

Di – di – di – dit

HAAT – Height above average terrain. A designation of elevation often used with antennas installed on very tall towers.

half-duplex - A communications mode in which a radio transmits and receives on two different frequencies but performs only one of these operations at a time. An operator using a half-duplex system would transmit and then listen for the other station's transmission. See: **duplex, full duplex, HDX**

half-wave antenna - An antenna consisting of a length of wire or conductive metal tubing that is electrically one-half wavelength long for the desired operating frequency. See: **resonance, wavelength**

Hallicrafters – A former manufacturer of Amateur Radio equipment.

ham (*slang*) – An Amateur Radio operator. Though the origin of the term is not certain, it is believed telegraphers in the early days of the use of Morse code called newcomers "hams."

ham-band-only receiver--A receiver which will only tune the bands assigned to or used by Ham radio operators. See: **general-coverage receiver**

ham bands – The frequencies on which Amateur Radio operations are authorized by the various agencies in countries around the world.

Ham City – A vendor of Amateur Radio equipment and supplies. Visit: **http://www.hamcity.com/**

hamfest – A gathering of Amateur Radio enthusiasts at which Hams meet to buy, sell, and swap equipment, visit with each other, and attend seminars on various subjects of interest to hobbyists. See: **boneyard, flea market**

Hammarlund – A former manufacturer of Amateur Radio equipment, best known for its line of receivers.

Ham Nation – A weekly television show dealing with Amateur Radio, streamed live via the Internet. Past shows are archived as well and available for viewing. Visit: **https://twit.tv/shows/ham-nation**

ham operator – A person who holds a valid Amateur Radio operator's license from his or her country's communications governing agency.

Ham Pros – A group of independent vendors of Amateur Radio equipment and supplies spread around the country including Associated Radio, Lentini Communications, Radio City, and Universal Radio. Visit: **http://www.hampros.com/**

Ham Radio Deluxe - A commercially available software system for Amateur Radio Operators which offers computer control of most transceivers, an electronic logbook, and ability to operate most digital modes. Visit: **http://www.ham-radio-deluxe.com/**

Ham Radio Magazine (*antiquated term*) – Amateur Radio magazine published from 1968 until 1990. Known for its emphasis on more technically-oriented content.

Ham Radio Outlet – A multi-location vendor that sells Amateur Radio equipment and supplies. Abbreviated as HRO. Visit: **http://www.hamradio.com/**

ham shack (*slang*) – The area in which a Ham has set up his radio station. It can be a separate building but is usually just a corner in a room, garage, or basement area.

Ham Station, The – A vendor of Amateur Radio equipment and supplies. Visit: **http://www.hamstation.com/**

hand-held (*slang*) - A small, battery-powered transceiver, usually for the VHF and/or UHF frequencies, so named because it is small enough to be carried easily in one hand. Sometimes called a brick, walkie-talkie or HT.

HandiHam System – An organization for Amateur Radio operators with physical disabilities. Visit: **http://www.handiham.org/**

handi-talkie (*slang*) - A small, hand-held battery-powered transceiver, usually for the VHF and/or UHF frequencies. Sometimes called a brick, walkie-talkie or HT.

handle (*slang*) - A Ham Radio operator's name. Most now prefer that "name" be used instead of handle or the Citizens Band terms "personal" or "first personal." On CW, the abbreviation "OP" is typically used instead of "name" or "handle."

hang time (*slang*) – The length of time following the end of a transmission on a repeater station before the repeater carrier drops. This short pause allows others who want to access the repeater a chance to do so before the signal drops or another station transmits. On many repeaters, a courtesy beep will alert users when the repeater is ready to accept another transmission.

harmonic – 1. The multiple of a fundamental frequency.

2. An undesirable spurious emission that occurs on a multiple of the frequency of the original transmission. This is a violation of rules and is even more serious if the harmonic transmission falls outside an Amateur Radio band, possibly causing interference to other services. See: **spurious emission**

2. (*slang*) The children of an Amateur Radio operator.

HDX - Half-duplex. See: **half duplex**

headphones – A small speaker or two small speakers mounted on a band that can be worn on the head with the speaker(s) on each ear. The speaker lead(s) is then plugged into the headphones outlet on the receiver/transceiver. See: **headset**

headset – Headphones with the addition of a small microphone on a boom so that it can be positioned near the lips. Sometimes called a boom-set. See: **boom-set, headphones**

health and welfare traffic – Messages originating from a disaster area that deal with the well-being of citizens in the affected area, intended for family or friends who might be concerned about them. Health and welfare messages have lower precedence than emergency or priority traffic and must wait until that traffic has been completed. See: **emergency traffic, precedence, priority traffic**

Heathkit – At one time, one of the leading manufacturers of Amateur Radio equipment, with most products offered in kit form.

Heil – A manufacturer of equipment for Hams, best known for its line of microphones and headsets/headphones. Founded by Bob Heil K9EID, a member of the Rock & Roll Hall of Fame for his work with sound techniques in concert venues. Visit: **http://www.heilsound.com/amateur/**

hellschreiber - A digital mode for sending and receiving text on the Amateur Radio HF bands. Often shortened to "hell."

henry – Unit of electrical inductance. See: **inductance**

hertz – The unit for measuring frequency. One hertz is one cycle per second of a repetitive wave, and typically refers to a sound or electromagnetic wave. One kilohertz (kHz) is 1,000 cycles per second. One megahertz (mHz) is one million cycles per second. One gigahertz is one billion cycles per second. Abbreviated as Hz.

heterodyne – 1. (*verb*) Putting two signals of differing frequencies together—intentionally or not—to create a third signal that is an arithmetical factor of the first two. Example: A signal of 3.700 megahertz mixed with a signal of 3.701 megahertz creates and audio signal of one kilohertz. This may be done inside a receiver to change the frequency of a signal to another frequency that may be more easily processed, or to create a tone so CW or SSB may be copied. See: **BFO, integral frequency**

2. (*noun*) A signal created by the mixing of two other signals of different frequencies. Such a signal can be the cause of interference.

hex beam – A popular Amateur Radio two-element wire beam antenna. The antenna uses wire for its elements and supporting spreaders are made of Fiberglas, bamboo or other material to form a two-element directive antenna. This method allows for the beam to have elements covering six through twenty meters all fed with a single coax feedline. Visit: **http://www.karinya.net/g3txq/hexbeam/**

HF – High frequency. See: **high frequency**

HH (*CW abbrev.*) – Error while sending, sent as one character, di-di-di-di-di-di-di-dit.

HI (*CW abbrev.*) – 1. Indicates the operator is laughing. Sometimes used on phone modes as well but generally frowned upon. The operator can simply laugh rather than saying, "Hi!" to indicate amusement.

2. (*CW abbrev.*) – "High."

hi fi audio – See: **extended single-sideband**

high frequency - Typically defined as that portion of the radio-frequency spectrum that falls between 3 and 30 megahertz. This range is also sometimes referred to as "shortwave." Abbreviated as HF. See: **EHF, shortwave, UHF, VHF, VLF**

high-pass filter - A filter designed to pass high frequency signals while blocking lower frequency signals. See: **filter**

holiday style (*slang*) – A Ham operating from a foreign country while on vacation or for work purposes and getting on the air casually as he or she has time or opportunity. Differs from a DXpedition in which operators attempt to be on the air around the clock. See: **DXpedition**

hollow state (*slang*) – Term for equipment that uses vacuum tubes, as opposed to solid state devices.

homebrew (*slang*) – 1. (*noun*) Radio equipment constructed by an individual, not purchased already built.
2. (*verb*) To construct a piece of equipment rather than purchase it commercially built.

home QTH (*slang*) – An operator's city or place of residence. Considered unnecessarily wordy and not encouraged. Comes from the Q-signal for, "What is your location?" Example: "The home QTH here is Springfield, Illinois," or, "I have arrived at the home QTH so I'll sign off now." See: **Q-signals, QTH**

hop (*slang*) – The distance between two stations communicating by reflecting the radio waves off of the ionosphere or some other object or entity. See: **multi-hop**

horizontal loop – A loop antenna that is parallel to the ground. Sometimes called a skywire loop. See: **loop, skywire**

horizontal polarization – 1. An electromagnetic wave which has its electrical lines of force parallel to the ground.
2. An antenna that has its element or elements roughly parallel to the ground beneath it. See: **vertical polarization**

HPE (*CW abbrev.*) – "Hope."

HQ 1. (*CW abbrev.*) – "Headquarters."
2. Abbreviation for American Radio Relay League headquarters in Newington, Connecticut. See: **American Radio Relay League**

HR (*CW abbrev.*) – "Here," "Hear."

HRD 1. (*CW abbrev.*) – "Heard."
2. Ham Radio Deluxe. See: **Ham Radio Deluxe**

HT – Handi-talkie. See: **hand-held, handi-talkie**

Hurricane Watch Net – An Amateur Radio network that goes into continuous session when a hurricane threatens landfall anywhere in North and Central America. Abbreviated as HWN. Visit: **http://www.hwn.org/**

HV 1. (*CW abbrev.*) – "Have."
2. Abbreviation for "high voltage."

HW (*CW abbrev.*) – "How."

HW?, HW NW? (*CW abbrev.*) (*slang*) - "How do you copy my signal now?" Or, "Back to you to see if you still copy my signal."

hybrid rig (*slang*) – A receiver or transceiver that uses both tubes and transistors in its various circuits.

Hz – Abbreviation for the unit of frequency, hertz. See: **hertz**

I

India

Di – dit

I - Symbol for electrical current in a circuit, measured in amperes.

iambic keying - A method of sending Morse code in which the operator closes both keyer paddle connections at the same time to send alternating dots and dashes. An electronic keyer with the proper logic circuitry and an iambic paddle are necessary to do this.

IARU - International Amateur Radio Union. See: **International Amateur Radio Union**

IC - Integrated circuit. A single device that contains several or many other types of electronic circuits inside it. See: **integrated circuit**

Icom - A major manufacturer of Amateur Radio equipment, headquartered in Japan. Visit: **http://www.icomamerica.com/**

ID – (*verb*) To identify a station that is transmitting, typically by giving the assigned call letters. See: **call letters, identify**
(*noun*) A station's call letters as assigned by the pertinent regulatory agency. See: **call letters, call sign**

identify – Giving the assigned call letters of the station at the beginning and end of a transmission or series of transmissions or at least every ten minutes during an ongoing conversation. FCC rules are specific about this operating requirement.

IF - Intermediate frequency. See: **intermediate frequency**

IMD – Intermodulation distortion.   Spurious emissions that can occur when two or more signals of different frequencies are mixed together in a receiver or transmitter creating additional signals that can create minor to severe interference within the receiver or transmitter.

impedance – Resistance to the flow of electric current, measured in ohms. Impedance is represented by the letter Z.

inductance - a measure of the ability of a coil to store energy in the form of a magnetic field. Unit of measurement is the henry.

inductor - an electrical component usually composed of a coil of wire wound on a central core. An inductor stores energy in a magnetic field. See: **coil, inductance**

informal traffic – A message or communication passed along from one station to another in a conversational manner. See: **formal traffic, net, precedence**

input, input frequency - the frequency on which a repeater station's receiver is set to listen for signals. This should be the transmit frequency for the operator's transmitter if he or she wishes to communicate through that repeater.

insulator - A material that effectively resists the flow of electric current.

integrated circuit - A single component that contains multiple devices. Abbreviated as IC.

intermediate frequency - A frequency to which a received signal is shifted from its original frequency as an intermediate step in processing that signal.   Converting incoming signals to an intermediate frequency can enhance amplification, filtering and signal processing. Abbreviated as IF. See: **heterodyne**

intermod – Intermodulation. When spurious signals are produced by two or more signals inadvertently mixing inside a receiver, such as at a repeater station where multiple transmitters may be located nearby. This can make it difficult for the station receiver to process incoming signals.

intermodulation distortion - Spurious emissions that can occur when two or more signals of different frequencies are mixed together in a receiver creating additional signals that can create minor to severe interference within the receiver.

internal tuner – An antenna matching device that is built into a transceiver. Such devices automatically attempt to find the best match to the antenna system when they are engaged, usually by a button push. Most commercially available Amateur Radio transceivers have an internal "tuner" or offer one as an option. See: **antenna tuner, auto-tuner**

International Amateur Radio Union - A worldwide Amateur Radio organization whose members consist of the official radio societies from all participating countries. In the USA, the ARRL is the country's member organization. Abbreviated as IARU. Visit: **http://www.iaru.org/**

International Morse code – The most accepted version of a method of transmitting text information as a series of dots and dashes sent and received as on-off tones, lights, or clicks that can be directly understood by a skilled listener or observer. The length of a dot is one unit and a dash is three units. Except with the Farnsworth method, the space between each part of a character is one unit, between each letter/number is three units, and between words is seven units. Today, and in Amateur Radio, the International Morse code is the standard version in use. Morse code was named for its inventor, Samuel F.B. Morse. In Amateur Radio, the mode is also often referred to as CW. See: **CW, code**

(Note: The header for each letter in this dictionary contains the "sound" of the appropriate letter in International Morse code, not dots and dashes. It is far better for someone desiring to learn the code to hear the way letters sound instead of learning the characters as dots and dashes. The brain has to make one additional translation from visual to sound from dots/dashes, slowing down the translation.)

International Telecommunications Union - The international body charged with specifying by treaty the worldwide guidelines concerning the use of the electromagnetic spectrum for communications purposes. Visit: **http://www.itu.int/**

Internet Radio Linking Project –A system that links individual amateur radio stations and repeater stations via the Internet using voice-over-IP (VoIP) software and hardware. Abbreviated as IRLP. Visit: **http://www.irlp.net/**

inverted V antenna - A wire dipole antenna in which the center is supported at a point that is higher than the two ends, forming an upside-down vee. Useful because it takes less space than a typical dipole. See: **dipole**

inverter - An electrical device that converts direct current (DC) to alternating current (AC).

I/O - Input/output.

ionosphere - The electrically charged region of the Earth's atmosphere that can refract radio signals. This region is located approximately 40 to 400 miles above the Earth's surface.

IOTA - Islands on the Air. See: **Islands on the Air**

IRC - International Reply Coupon. A coupon that can be purchased at post offices throughout the world that can be exchanged in foreign countries for return postage for a surface mail letter to the country that issued the coupon. Use and recognition of IRCs are rapidly diminishing.

IRLP – Internet Radio Linking Project. See: **Internet Radio Linking Project**

Islands on the Air - A movement to encourage Amateur Radio operation from and with islands throughout the world. Abbreviated as IOTA. Visit: **http://www.rsgbiota.org**

isotropic – A theoretical single point source of radiated radio frequency energy used to calculate and compare gain of various antennas.

ITU - International Telecommunications Union. See: **International Telecommunications Union**

J

Juliet ("Jew – lee – ETT")

Di – dah – dah – dah

J-38 (*antiquated term*) – Now an antique, a type of simple but sturdy Morse code straight key originally manufactured for railroad telegraph and military use. See: **key**, **straight key**

jack – A female electrical connector designed to accept a plug. See: **plug**

jam (*slang*) – To deliberately cause intentional interference by transmitting on a frequency already in use by other stations. Such operation is illegal in the Amateur Radio service and can lead to monetary fines and jail time.

Jamboree on the Air - An annual on-the-air event in which Boy Scouts worldwide attempt to make contact with each other using Amateur Radio. Abbreviated as JOTA.

Johnson, E. F. – A former manufacturer of Amateur Radio equipment, best known for their series of Viking transmitters.

JOTA - Jamboree on the Air. See: **Jamboree on the Air**

J-pole – a vertical antenna consisting of a half-wavelength radiator fed by a quarter-wave matching stub. The physical shape resembles the letter "J," thus the name.

JT-65 – One of the digital operating modes available using WSJT software. JT-65 is especially useful when employed for weak-signal radio communication between Amateur Radio operators. The mode is named after Joe Taylor, K1JT, the Nobel laureate who developed the mode. See: **digital**, **digital modes**, **WSJT**

jug (*slang*) – Very large transmitting tubes.

juice (*antiquated term*) (*slang*) – Early term for electrical current.

jumper – 1. A small piece of wire used to electrically connect two different points in a circuit.
2. A short piece of coax cable used to connect transmitters, receivers, and/or transceivers to other station accessories such as meters, tuners, or antenna switches.

junkbox (*slang*) - A collection of spare parts and miscellaneous items retained—not necessarily in a single container—by Hams, makers, or other electronic hobbyists.

jury rig (*slang*) - Fix a problem in a sloppy or unorthodox way.

K

Kilo ("Kee – low")

Dah – di – dah

K (*CW abbrev.*) – "I am finished with my transmission. Go ahead."

K9YA Telegraph – An e-zine for the amateur radio community, offered free on-line in a PDF form. Visit: **http://www.k9ya.org/**

KC (*antiquated term*) – Abbreviation for 1000 cycles per second or kilocycles. Replaced by kilohertz or kHz. See: **hertz**

Kenwood - A major manufacturer of Amateur Radio equipment, headquartered in Japan. Visit: **http://www.kenwoodusa.com/**

kerchunker (*slang*) – A person who transmits to activate a repeater station but does not identify his or her station. Such practice is generally discouraged and is technically illegal since the operator does not identify. See: **kerchunking**

kerchunking (*slang*) - Activating a repeater station by making very short transmissions without identifying. Stations sometimes do this to determine if they have a strong enough signal to work through the repeater or not. Those who do are called kerchunkers.

key – 1. (*noun*) A switch or button used to create dots and dashes (dahs and dits) in Morse code.
2. (*verb*) To use such a device to create the dots and dashes and form characters.
3. (*verb*) To engage the push-to-talk switch on a microphone to begin transmitting. See: **key-up, un-key**

---

key clicks - Undesired "clicks" or "thumps" generated by a CW transmitter as the Morse code key is used to turn the carrier on and off. Clicks can cause interference to other stations operating on the band. See: **clicks**

keyer – An electronic device for creating the dots and dashes of Morse code semi-automatically using a key that is commonly referred to as a "paddle." Dots are made by pressing the paddle one way and dashes by pressing the paddle the other. See: **electronic keyer**, **key**, **paddle**

key-up - Activating a repeater station by transmitting on its input frequency. Example: "I was able to key up the 88 repeater from my office downtown."

Kids Day – A twice-a-year operating event in which young Hams and those interested in the hobby are urged to operate a station and talk with each other as well as with other Amateurs. Fathers and mothers are also encouraged to allow their own children to operate. Sponsored by the American Radio Relay League. Visit: **http://www.arrl.org/kids-day**

kilo - The metric prefix meaning multiply the suffix value by 1000.

kilocycle (*antiquated term*) – A frequency of one thousand cycles per second. This term has been replaced by kilohertz (kHz). See: **hertz**

kilohertz – A frequency of one thousand hertz. See: **hertz**

kilowatts – One thousand watts. See: **watt**

K-index - A measure of the Earth's magnetic field. Lower numbers typically mean better propagation of radio waves.

KN (*CW abbrev.*) - "I am finished with my transmission. The station to which I am talking may go ahead but no breakers, please." Sent as a single character, dah-di-dah-dah-dit. See: **breaker**

L

Lima ("LEE – muh")

Di – dah – di - dit

L – Abbreviation for the electrical term inductance.

ladder line –Balanced, two-conductor transmission line, typically with an impedance between 300 and 600 ohms. The two conductors are separated uniformly by some non-conductive spacing material placed at regular intervals. See: **balanced line, open wire line, window line**

landline (*slang*) (*antiquated term*) – The common telephone. The term is used to make it clear the operator is not talking about the phone mode of transmission. Example: "If we lose each other in the static, give me a call on the landline." See: **twisted pair**

LCD - Liquid crystal display. See: **liquid crystal display**

LDG Electronics – Manufacturer of Amateur Radio equipment. Visit: **www.ldgelectronics.com**

LDE (*slang*) – Long delayed echo. See: **long delayed echo**

lead – (pronounced "leed") 1. A wire or connection point on an electrical component.
2. A probe and the wire attached to it that is used to establish a connection from a test instrument to the point or points in a device where a parameter is going to be measured.

League, The (*slang*) – The American Radio Relay League. See: **American Radio Relay League**

LED - Light-emitting diode. See: **light emitting diode**

LF - Low frequency. See: **low frequency**

Lid (*slang*) – A poor operator.

light emitting diode - A two-lead semiconductor which emits light when activated. Often used as a light source or visual indicator. Abbreviated as LED.

lightning arrestor - A device used to help protect structures, power lines, and equipment from lightning damage by shunting some of the energy to an earth ground system.

linear (*slang*) – An external power amplifier. See: **amp, amplifier, barefoot, linear amplifier**

linear amplifier - An external power amplifier used after the transceiver output. Linear means that the signal emitted by the amplifier is directly proportional to the signal that goes in. See: **amp, amplifier, barefoot, linear**

linear power supply – A device that converts 110-volt alternating current (AC) into nominal 12-volt (typically 13.8 volts) direct current (DC). A steel or iron laminated transformer reduces the input current that is then rectified by diodes and smoothed into low voltage DC by electrolytic capacitors. See: **power supply, switching power supply**

line-of-sight - A form of radio propagation in which the emitted signal travels a straight-line path directly from one station's antenna to the other. This means for good communication, each antenna should be within sight of the other. See: **radio horizon**

liquid crystal display – A type of visual display using two sheets of material with a liquid crystal solution between them. When an electric current is passed through the liquid, the crystals act like a shutter, allowing light to pass through or blocking it. Abbreviated as LCD. Such technology is now often used in Amateur Radio transceiver displays.

little pistol (*slang*) – A modest Amateur Radio station setup with inexpensive, lower-power equipment and limited antennas. See: **big gun**, **peanut whistle**

LMR-200, LMR-400 - Common types of coaxial cable used by Hams as feedlines for antenna systems. See: **RG-6, RG-8, RG-8X, RG-58, RG-59, 9913**

load - An electrical component or portion of a circuit that consumes power.

lobe – The area in the radiation pattern of an antenna in which the signal strength is at its maximum. See: **null**

log – 1. (*noun*) A document maintained by Hams that lists the details of their stations' contacts as well as other details of operation of the station. Though no longer required by the Federal Communications Commission, such records can come in handy in the case of interference claims and to track propagation conditions. A log is also necessary for claiming a score in most radiosport events.
2. (*verb*) The process of maintaining a log. Many Hams now use computer logging. See: **logging software**

logger (*slang*) – An individual who helps a contest operator by logging contacts and their details.

log periodic antenna – A multi-element beam antenna that is characteristically very broad-banded due to the specifically calculated RF energy interaction among its elements. See: **beam, element**

Logbook of the World – An Internet service on which stations may log contacts for the purpose of applying for the operating awards sponsored by the American Radio Relay League as well as some by *CQ Magazine*. Abbreviated as LOTW. Visit: **http://www.arrl.org/logbook-of-the-world**

logging software – Computer programs that allow Hams to log details of contacts and maintain a database of on-air activity electronically. May be used regularly and/or for radiosport events.

lollipop (*slang*) (*antiquated term*) - Nickname for the distinctively-shaped Astatic D-104 microphone.

long delayed echo (*slang*) – A signal that is heard seconds or even minutes after it was actually transmitted, often by the station that sent it out in the first place, due to the long route it may have taken. The route may have been all the way around the planet. Abbreviated as LDE.

long path – A signal path that is a reciprocal of the shortest route from transmitting station to receiving station. See: **great circle bearing, short path**

loop – An antenna whose radiating element is one continuous conductor, with the feedline connected to each end. The loop is typically hung in a circle, in a square or rectangle, or as a triangle, as with a delta loop. See: **delta loop, horizontal loop, magnetic loop**

LOTW – Logbook of the World. See: **Logbook of the World**

lower side-band, lower sideband - The frequencies on a carrier that are lower than the carrier frequency, but that contain power as a result of the modulation process. Operators using single-sideband can choose to operate either lower or upper sideband. Typically and by convention, LSB is used on 160, 80, 60 and 40 meters. USB (upper sideband) is used on all other bands. Abbreviated as LSB. See: **LSB, upper side-band, USB**

lowest usable frequency - The lowest frequency that can support reliable propagation between stations. Abbreviated as LUF.

low frequency - The radio spectrum between 30 and 300 kilohertz. Abbreviated as LF.

low-pass filter - A filter that allows signals below the cutoff frequency to pass through and reduces signals above the cutoff frequency. See: **high-pass filter**

LSB - Lower side-band. See: **lower sideband**

LTR (*CW abbrev.*) – "Later," "Letter."

LUF – Lowest usable frequency. See: **lowest usable frequency**

LV (*CW abbrev.*) – "Leave," "Love."

LVG (*CW abbrev.*) – "Leaving," "Loving."

M

Mike

Dah – dah

M – 1. Abbreviation for mega or one million. Example: "2 mHz" is two megahertz, or two million cycles per second.

2. Abbreviation for meter, a metric unit of measure of distance. Example: A "20M beam" is a 20-meter beam.

machine (*slang*) - A repeater station. Example: "Let's switch over to the Paducah machine."

MacLoggerDX – A computer software program for logging Amateur Radio contacts and for use in radiosport events. This system is specifically designed for use with Apple computers and the OSX operating system. Visit: **http://www.dogparksoftware.com/MacLoggerDX.html**

magic band (*slang*) – A term used for the 6-meter Amateur Radio band because of its unusual and interesting propagation characteristics.

mag-mount (*slang*) - An antenna with a magnetic base for mounting purposes. This allows quick installation and removal from a motor vehicle or other metal surface. Used primarily for VHF and UHF antennas since such a mount would not be reliable for longer ones.

magnetic loop – A very small but very narrow-banded loop antenna.

magnetic mount – See: **mag-mount**

Maidenhead locator system - A geographic coordinate system of grid squares that divide up maps of the Earth. These grid squares are used by Amateur Radio operators to determine a contact's location. As a personal challenge, many operators attempt to contact stations in as many grid squares as possible. The system's name comes from the English town where it was first created by VHF enthusiasts. See: **grid square**

Main Trading Company – A vendor of Amateur Radio equipment and supplies. Abbreviated as MTC. Visit: **http://www.mtcradio.com/**

maker – One who participates in the "maker movement." Many Hams find themselves right at home in the community of "makers" and many people who are already active in the movement are becoming interested in Amateur Radio. Visit: **http://makerfaire.com/** See: **maker faire**, **maker movement**

maker faire – A gathering of tech enthusiasts, crafters, and hobbyists, who are interested in designing and building things of all types. A part of the "maker" movement. Amateur Radio has become a key element in many maker faires. Visit: **http://makerfaire.com/** See: **maker**, **maker movement**

maker movement - A community of creative and curious people, including hobbyists, enthusiasts and students who are interested in building from scratch all types of things while employing emerging technology. Visit: **http://makerfaire.com/maker-movement/** See: **maker**, **maker faire**

making the trip (*slang*) - Confirmation that a received signal is readable. Example: "You're making the trip to the repeater just fine today."

malicious interference – Deliberately and intentionally causing interference to another radio transmission. This is not only bad operating practice but illegal and can result in a fine or incarceration.

maritime mobile – An Amateur Radio station operating aboard a maritime vessel. Technically such operation requires that the station be outside the territorial waters of any nation and the operator must follow the rules and regulations of the country under which the vessel is flagged. On CW, /MM should be sent after the call sign of the operator.

MARS – Military Affiliate Radio System. See: **Military Affiliate Radio System**

match (*slang*) – 1. (*noun*) A condition in which the output of a transmitter is the same or nearly the same impedance as the antenna system in use, thus offering the most efficient transfer of power from the rig to the load (antenna). A 50-ohm output to a 50-ohm antenna system is a match. See: **mismatch**
 2. (*verb*) The process of finding a point at which the impedance of the antenna system is the same or nearly the same as the output of the transmitter, when capacitive and inductive reactance cancel out leaving primarily radiation resistance and maximum radiation of RF energy. This may already be the case but may also involve using a matching device such as an "antenna tuner." See: **antenna tuner, resonance**

matchbox (*slang*) – A term for an antenna-impedance matching device, commonly called an "antenna tuner." See: **antenna tuner, match**

matching stub - A length of transmission line that is used to help bring an antenna system into resonance. By choosing the proper length and characteristic impedance of line, and having one end open or shorted, a stub becomes in effect a capacitor or inductor and can be used to achieve a match when inserted at a selected point in the regular transmission line. See: **feed line, match, resonance**

maximum usable frequency - The highest radio frequency that can be reliably used for transmission between two points by way of reflection from the ionosphere. Abbreviated as MUF.

mayday – When transmitted over the air, indicates that a life-threatening event is being reported or relayed. This distress call is typically used on voice transmissions as opposed to "SOS" on Morse code or digital modes. See: **distress call, SOS**

MC (*antiquated term*) – Megacycle. One million cycles per second. Now usually replaced by the term "megahertz." See: **hertz**

medium frequency - The radio-frequency spectrum from 300 to 3,000 kilohertz. Sometimes called "medium wave." Abbreviated as MF.

medium wave - The portion of the radio-frequency spectrum from 300 to 3000 kilohertz. Abbreviated as MW. The term is also commonly used to refer to the AM commercial broadcast band, 535 to 1705 kilohertz.

mega - The metric prefix meaning multiply the suffix value by 1,000,000.

megacycle (*antiquated term*) - One million cycles per second, abbreviated MC. Now usually replaced by megahertz. See: **hertz**

megahertz – One million hertz or cycles per second in a recurring wave. See: **hertz**

memories – Programmable settings in a receiver, scanner or transceiver that allow the operator to enter frequencies or channels that are often used so they can be more quickly accessed or scanned. See: **scanner**

memory channel – Frequency, mode, and other information stored by a radio so it can be easily chosen by the operator or scanned by the radio. See: **memories**

memory effect – A phenomenon with certain types of rechargeable batteries in which, if they are not completely discharged before being recharged, they will begin to lose capacity and have to be recharged more often.

meteor scatter – The use of the ionized trails of meteors in or near the Earth's atmosphere to reflect radio signals back to other stations.

Mexican Federation of Radioexperimentadores – The national organization for Amateur Radio Operators in Mexico. Visit at **http://www.fmre.org.mx/** See: **FMRE**

MF - Medium frequency. See: **medium frequency**

MFJ – A major manufacturer of a wide array of Amateur Radio equipment and accessories, headquartered in Starkville, Mississippi. Visit: **http://www.mfjenterprises.com/**

mic (slang) – Microphone. A device that converts audio into electrical energy. Sometimes spelled "mike."

mickey mouse (*slang*) – Spoken by an operator who is operating maritime mobile, taken from the Morse code identifier for such portable operations, "MM." See: **maritime mobile**

micro - The metric prefix meaning divide the suffix value by 1,000,000.

microphone – A device that converts audio into electrical energy. See: **mic, mike, PTT**

microphone gain – 1. The sensitivity of a microphone amplifier in a transmitter or transceiver.
2. The control that allows adjustment of a microphone amplifier for more or less gain.

microphone to you (*slang*) – "I have completed my transmission and now it is your turn to talk." Example: "That's all I have for now. Microphone to you, Jack."

microwave – Typically defined as the region of the radio-frequency spectrum above one gigahertz. See: **hertz**

mike (*slang*) – Microphone. See: **microphone**

Military Affiliate Radio System - A civilian auxiliary service sponsored by the Department of Defense in the USA consisting primarily of licensed Amateur Radio Operators who are interested in assisting the military with communications on a local, national, and international basis as an adjunct to normal communications. When participating in MARS activities, stations operate on military frequencies outside the usual Amateur bands. MARS stations must also be licensed by the military in addition to their FCC license and are assigned different call signs than their FCC-issued Amateur call letters. Visit: **http://www.netcom.army.mil/mars**

mill (*antiquated term*) – A typewriter designed to be used by telegraph operators to copy Morse code.

milli - The metric prefix meaning divide the suffix value by 1000.

mismatch (*slang*) - A condition in which the output of a transmitter is not nearly the same impedance as the antenna system in use, thus offering an inefficient transfer of power from the rig to the load. A mismatch that is nearer to the needed impedance may be achieved by using a matching device, sometimes called a "tuner" or matchbox. See: **match, matchbox, tuner**

mixer - A receiver circuit that takes two or more input signals then produces an output that includes the sum and difference of those signals' frequencies. This creates a frequency at which the signal can more easily be processed. See: **heterodyne**

MM – When sent in Morse code following an Amateur Radio call sign, the indication that the station is operating maritime mobile. See: **maritime mobile**

MNI (*CW abbrev.*) – "Many."

mobile – 1. An Amateur Radio station installed in a vehicle or a portable transceiver that can be used in a vehicle.

2. The process of operating an Amateur Radio station from a vehicle of any type. See: **base station, fixed station, portable**

3. (*slang*) Indication by an operator that he is operating mobile. Example: "This is K2XYZ mobile in North Carolina."

mode - The specific type of electromagnetic signal being generated by a transmitting device. Examples: AM, FM, or single-sideband. Emission modes are more completely described and defined by regulatory agencies in each country, such as the Federal Communications Commission in the USA. See: **emission mode**

modem - Modulator/demodulator. A circuit or device that modulates an audio or radio signal to transmit data and demodulates a received signal to recover that transmitted data.

modulate – To insert a signal containing information such as audio onto a higher-frequency wave such as a radio-frequency carrier. See: **AM, amplitude modulation, FM, frequency modulation**

modulation - Process by which a signal containing information such as audio is used to modify a higher-frequency wave such as a radio-frequency carrier. See: **AM, amplitude modulation, FM, frequency modulation**

Molex connector - A nylon-body plug often used for power connections.

monitoring (*slang*) – 1. Listening to a radio or scanner. Example: "I'll be monitoring 52 simplex for your call."

2. An on-air indication that an operator is available and listening for anyone to call. Example: "This is K1QQQ monitoring. Anybody around?"

mono-band - A rig, device or antenna that can be used on only one band.

mono-pole – A single element vertical antenna.

moonbounce - See: **Earth-moon-Earth**

MORN (CW abbrev.) – "Morning."

Morse code – A method of transmitting text information as a series of dots and dashes sent and received as on-off tones, lights, or clicks that can be directly understood by a skilled listener or observer. Today, and in Amateur Radio, the International Morse code is the standard version in use. Morse code was named for its inventor, Samuel F.B. Morse. In Amateur Radio, the mode is also often referred to as CW. See the definition for International Morse code for an explanation for why I use dits and dahs rather than dots and dashes at the beginning of each section of this dictionary. See: **CW, code, International Morse code**

MOSFET - Metal-oxide semiconductor field-effect transistor.

motorboating (*slang*) – A fluttering, low-frequency noise on the audio of a transmitting station, so named because the sound resembles that of an outboard boat motor.

mount – The method by which an antenna is attached to a vehicle. See: **ball mount, bumper mount, fender mount, mag mount**

MSG (*CW abbrev.*) – "Message."

MUF - Maximum usable frequency. See: **maximum usable frequency**

muffin fan – A small electrically-powered fan typically used to circulate air through or over electronic equipment.

multi-band – A rig, device or antenna that can be used on more than one Amateur Radio band.

multi-hop - A radio signal that has been refracted by the ionosphere and the Earth two or more times between origination and eventual reception. See: **hop**

multimeter, multi-meter -- A test instrument used to measure various values in a circuit including current, voltage and resistance.

multi-mode transceiver – A transceiver that is capable of operating most emission modes available to Amateur Radio licensees, including single-sideband, CW, AM, and FM.

multi-path – Signals reaching a receiver by way of more than one path due to reflection or refraction off objects, hills, or the ionosphere. This occurs most often on VHF and UHF and can cause a signal to interfere with itself and be difficult to copy. See: **direct path**

multiplier – Criterion within radiosport that allows for greater scoring. Example: The number of individual countries worked in a contest may be multiplied by the total number of contacts, thus making a country a multiplier. If a station makes 500 contacts in 100 different countries in a contest, his score would be 50,000 points (500 X 100).

MW - Medium wave. See: **medium wave**

N

November

Dah – dit

N (*CW abbrev.*) – "No."

N1MM Logger – A popular computer software program for logging Amateur Radio contacts. This program is primarily used for radiosport logging and is available for free. Visit: **http://n1mm.hamdocs.com/tiki-index.php**

N3FJP Amateur Contact Log – An inexpensive and popular commercially available software program for logging Amateur Radio contacts. This program is primarily for day-to-logging use although the author also offers versions of the system for specific radiosport events. Visit: **http://www.n3fjp.com/**

NA (*CW abbrev.*) – "North America."

narrowband FM - A variation of the frequency modulation (FM) mode in which the modulation information on the carrier signal only deviates about 2.5 kilohertz above and below the center frequency. Abbreviated as **NBFM**.

National – A former manufacturer of Amateur Radio equipment, especially known for its line of receivers.

National Contest Journal - A magazine published six times per year by the American Radio Relay League. Covers topics of interest to Amateur Radio operators who enjoy radiosport. Abbreviated as NCJ. Visit: **http://ncjweb.com/**

National Electrical Code - A nationally-accepted set of electrical safety guidelines. Amateur Radio antennas are included. Abbreviated as NEC. Visit: **http://www.necconnect.org/**

National Institute of Standards and Technology - USA government agency that maintains official standards for the measure of such things as time and frequency. Abbreviated as NIST. Formerly known as The National Bureau of Standards. This agency operates WWV and WWVH radio stations that broadcast highly accurate time signals on precise shortwave frequencies. See: **WWV, WWVH**

National Radio Quiet Zone - An area in Maryland, Virginia, and West Virginia near the high-sensitivity radio telescopes operated by various branches of the government. There are varying restrictions on Amateur Radio transmissions—as well as those by other radio services—depending on how close the transmitters are to the protected facilities. Visit: **https://science.nrao.edu/facilities/gbt/interference-protection/nrqz/**

National Traffic System - A national system of Amateur Radio on-air networks designed to relay traffic via Ham Radio in the form of formal messages. Abbreviated as NTS. The NTS is administered by the American Radio Relay League. See: **traffic**

NATO phonetic alphabet – The generally accepted list of words that allows an operator on phone modes to spell out words to distinguish letters for clearer understanding. The heading for each letter in this dictionary shows the phonetic word for that character. See: **phonetic alphabet**

NB - Noise blanker. See: **noise blanker**

NBFM - Narrowband FM. See: **narrowband FM**

NCJ – National Contest Journal. See: **National Contest Journal**

N-connector - A threaded, weatherproof, medium-size RF connector used to join coaxial cables. Preferred use is in VHF and UHF frequency ranges.

NCS - Net control station. See: **net control station**

near field - The region of electromagnetic energy immediately surrounding an antenna.

near-vertical incidence skywave – A form of radio signal propagation in which the emitted wave goes up at a very sharp angle and is reflected back to Earth by the ionosphere in an area within a small radius—depending on the frequency being used—around the transmitting station. Such propagation is best employed for local and regional communication. Abbreviated as NVIS. See: **angle of radiation, critical angle, NVIS**

NEC – 1. National Electrical Code. See: **National Electrical Code**
2. Numerical Electromagnetics Code. See: **Numerical Electromagnetics Code**

negative (*slang*) – "No," "Incorrect."

negative copy (*slang*) – "Unable to copy any of your last transmission or message."

negative offset - A repeater's input frequency is lower than its output frequency. Example: A repeater that transmits at 146.88 megahertz but is listening for users at 146.28 megahertz is said to have a negative offset of 600 kilohertz. See: **positive offset**

net (*slang*) – Network. See: **network**

net control station - The station in control of a formal net who keeps the proceedings orderly and moving along. Abbreviated as NCS. See: **net, network**

network - A group of stations that meets on a specified frequency at a predetermined regular time. A net may be informal, formal or somewhere in between. A formal net is organized and directed by a net control station (NCS), who calls the net to order, recognizes stations as they enter and leave the net, and authorizes stations when they may transmit or begin to pass traffic. An informal net may allow participants to transmit at will in no particular order or as a "roundtable," in which stations take turns. Usually shortened to "net." See: **net control station, roundtable, traffic**

next over (*slang*) – The next time the station has a chance to transmit. Example: "John, please tell me your location on your next over."

NiCad - Nickel cadmium, pronounced "NYE – cad." The chemical composition in a popular type of rechargeable battery. The term usually refers to the battery itself. Example: "I have a bunch of NiCads charged and ready to go if we have a power failure."

nickels (*slang*) – A term often used on DX or county-hunting nets meaning a signal report of "55" or "Five by five." Origin: Citizens Band radio. See: **double nickels, RST, signal report**

NIL (CW abbrev.) – "None," "Nothing."

NiMH - Nickel metal hydride. The chemical composition in a popular type of rechargeable battery.

NIST – National Institute of Standards and Technology. See: **National Institute of Standards and Technology**

NMO - A type of screw-on mobile antenna coaxial mounting arrangement often used for VHF and UHF antennas.

node - A remotely controlled digipeater that is used as a connect point in packet radio. See: **digipeater, packet radio**

noise - Undesirable electromagnetic energy that causes interference with a signal's reception.

noise blanker - A circuit designed to reduce pulse-type noise in a receiver. Abbreviated as NB.

noise floor - The total noise in a receiver from all sources, internal and external. To be readable, a signal has to be greater than the noise floor. See: **noise**

noise reduction – A feature of digital signal processing used to reduce unwanted noise on a signal. Abbreviated as NR. See: **digital signal processing, noise**

notch filter – A narrow rejection filter for elimination of interfering signals. This type of filter often can have its rejection frequency varied with a control to be able to reduce an interfering signal wherever it may be above or below the frequency of the desired signal. See: **filter**

Novice (*antiquated term*) – An entry-level Amateur Radio license class in the USA that required a 5-word-per-minute Morse code examination and a basic theory exam. On-air privileges were very limited. Though the license class no longer exists, the Federal Communications Commission continues to renew those licenses that remain in effect.

NR – 1. Noise reduction. See: **noise reduction**
2. (*CW abbrev.*) – "Number."

NTS - National Traffic System. See: **National Traffic System**

null – 1. The area in the radiation pattern of an antenna in which the strength of a radiated signal is its lowest. See: **lobe**
2. To tune or adjust for a negative value.

number stations (*slang*) – Mysterious shortwave stations on which people can be heard reading what appear to be coded messages consisting of all numbers or letters.

Numerical Electromagnetics Code - A general-use software program for modeling antenna design and performance. Abbreviated as NEC. See: **antenna modeling, EZNEC**

NVIS - Near-vertical incidence skywave. See: **near-vertical incidence skywave**

NW (*CW abbrev.*) – "Now."

O

Oscar

Dah – dah – dah

OBS – Official Bulletin Station. See: **Official Bulletin Station**

OC (*CW abbrev.*) – Oceania, the part of the world that includes Australia, New Zealand, and the South Pacific islands.

OCF dipole – Off-center-fed dipole. See: **off-center-fed dipole**

odd split (*slang*) – An unconventional frequency separation between input and output frequencies of a repeater station. Most repeater stations in the USA use a 600 kilohertz split between input and output. See: **offset**, **split**

OES – Official Emergency Station. See: **Official Emergency Station**

off-center-fed dipole - A dipole antenna that is fed at a point away from its center, as with the traditional dipole design. The feedpoint is at a position nearer one end of an element where the impedance will be such that the antenna will work on multiple Amateur bands. Abbreviated OCF dipole. See: **dipole**.

Official Bulletin – A news story or informational article related to Amateur Radio or ARRL activities disseminated through regular multi-mode broadcasts from the American Radio Relay League station, W1AW, via email, by League appointees called Official Bulletin Stations, or other means. The stories are originated by the ARRL. See: **ARRL, American Radio Relay League, Official Bulletin Station, W1AW**

Official Bulletin Station – An appointed volunteer who receives and disseminates to other Amateurs and the general public Official Bulletins from the American Radio Relay League. Abbreviated as OBS. See: **ARRL**, **Official Bulletin**

Official Emergency Station - An appointee in the American Radio Relay League's Field Organization who is responsible for specific duties during drills or emergency situations. Abbreviated as OES. See: **ARRL**, **Field Organization**

Official Observer - A volunteer who monitors the Amateur Radio bands for rules violations or technical problems that may be causing interference and informs the stations involved. Abbreviated as OO. The OO program is administered by the American Radio Relay League.

offset (*slang*) – The difference between the input and output frequencies for a repeater station. To be able to listen and transmit at the same time, repeaters require two different frequencies. In the USA, the standard offset is 600 kilohertz. As a general rule, if the output (transmit) frequency of the repeater is below 147 megahertz then the input (listening) frequency is 600 kilohertz lower. This is referred to as a negative offset. If the output is above 147 megahertz, the input is 600 kilohertz above the output. This is termed a positive offset. See: **negative offset, odd split, positive offset, split**

offset frequency – See: **offset**

ohm - The fundamental unit of measurement of resistance as well as impedance.

Ohm's law – One of the most basic laws of electronics, this is a formula that shows the relationship between voltage (E), current (I) and resistance (R) in an electrical circuit. Ohm's law says that the voltage applied to a circuit is equal to the current flowing through the circuit times the resistance of the circuit, or the mathematical expression: $E = IR$, where E is voltage in volts, I is current in amperes and R is resistance in ohms.

old man (*slang*) – Term used to denote friendship between two male amateur radio operators, regardless their chronological ages. On Morse code, this is abbreviated "OM."

Olivia – An Amateur Radio digital mode designed to work in difficult communication conditions. It is not yet as popular as some other modes and is only available in a few of the commercially available software systems for digital modes. See: **digital modes**

OM (*CW abbrev.*) – "Old man." See: **old man**

omni-directional antenna – An antenna that receives or emits a signal near equally in all directions. See: **directional antenna**

one-by-one call sign - A special temporary call sign consisting of a single letter, a number, and another single letter, issued to special event stations or for other distinct purposes. Example: The call sign N9N was issued for the special event station commemorating the fiftieth anniversary of the journey to the North Pole by the submarine USS *Nautilus*. Visit: **http://www.1x1callsigns.org/**

one-way communication – A transmission on Amateur Radio bands that is not intended to be answered, beyond just simple testing. FCC rules place strict limitations on such transmissions and they are not generally authorized.

OO - Official Observer. See: **Official Observer**

OP (*CW abbrev.*) – "Operator," "Amateur Radio operator," "My name is…" See: **handle**

opening (*slang*) – A condition in which better than usual propagation on a particular band is occurring. Example: "There was a great opening on 6 meters this morning. I worked several Caribbean stations."

open repeater (*slang*) - A repeater whose access is available to any Amateur Radio operator who holds license privileges for its operating frequency. See: **closed repeater**

open wire line – A type of balanced antenna feedline that has two conductors spaced uniformly from one end to the other, usually by non-conductive spacers.   See: **balanced line**, **ladder line**, **window line**

operating event – An on-the-air event for Amateur Radio operators similar to a contest (radiosport) but entrants do not compete against each other for scores.   Instead they strive for personal achievement and individual goals. Examples: Field Day, Straight Key Night. See: **contest**, **Field Day**, **Straight Key Night**

OPR (*CW abbrev.*) – "Operate," "Operator," "Operating."

OSCAR - Orbiting satellite carrying Amateur Radio. A series of space satellites built, launched and controlled by Ham operators. Visit: **http://www.amsat.org/** See: **AMSAT**

oscillate – To vibrate. In electronics, this typically means to generate an alternating current, radio wave, or other periodic signal.

oscillator – A circuit within a transmitter that initially generates a radio frequency signal.

oscilloscope - an electronic test instrument used to observe wave forms on a display screen.

OT (*CW abbrev.*) (*slang*) – "Old timer." Someone who has been in the hobby for a long time.

outgoing QSL bureau – A system run by the American Radio Relay League to send out QSL cards for Hams in the USA to QSL bureaus in other countries. Cards can only go to countries with QSL bureaus that have a working relationship with the ARRL. Visit: **http://www.arrl.org/outgoing-qsl-service**   See: **QSL**, **QSL bureau**, **QSL card**

output frequency - The frequency of a repeater station's transmitter, and the frequency to which a user should have his receiver tuned.

over (*slang*) – 1. When used during a two way communication, it means the transmitting station has completed the transmission and is telling the station to which he or she is talking that it is okay to begin to transmit. Use of the term "over" is not recommended on repeater stations or if propagation conditions are good.

2. An opportunity to make a transmission. Example: "Bob, tell me what kind of rig you are running on your next over."

overload – When a received signal is so strong that it overcomes the normal operating parameters of the receiver's circuits creating undesirable images and false signals at various frequencies.

OVR (*CW abbrev.*) – "Over."

P

Papa ("Puh – PAH")

Di – dah – dah – dit

PA – Power amplifier.

packet - A unit of data sent across a network. In Amateur Radio, packets are sent via the airwaves.

packet cluster - A network of automated Amateur Radio packet stations that communicate information about DX stations that have been spotted on the air, contest reports, and other information.

packet radio - A system of digital communication in which information is transmitted over the air in short bursts. The bursts—or "packets"—usually contain the sending station's call sign along with other information. A terminal node controller (TNC) is used to encode and decode the information for broadcast and reception. Visit: **https://www.tapr.org/** See: **digi-peater, node, terminal node controller**

paddle - a key for sending Morse code with an electronic keyer. Dots to form Morse code characters are made by pressing the paddle one way and dashes by pressing the paddle the other. See: **electronic keyer, keyer**

panadapter – A scope or screen display used to visually monitor a wide band of frequencies at the same time. Sometimes called a "spectrum scope." See: **spectrum scope**

parallel conductor feed line – Antenna feed line constructed of two conductors held apart at a constant distance. May be encased in plastic to keep that distance consistent or constructed with insulating spacers placed at intervals along the line.

parasitic – 1. Undesired oscillations in a transmitter that create spurious signals on frequencies other than the desired one.

2. Interaction between elements in a beam antenna. See: **beam, parasitic element**

parasitic element - An antenna element that has effect on how the antenna performs by re-radiating or reflecting RF from the driven element. A parasitic element is not electrically connected directly to the feed line or to any other part of the antenna but, in the case of a beam, is connected mechanically to the boom. See: **beam, boom, parasitic**

Part 97 – The section of the rules and regulations of the Federal Communications Commission that deals with the Amateur Radio Service. See: **Amateur Radio, Amateur Radio Service, FCC, Federal Communications Commission**

pass (*slang*) - The time period when the signals from an orbiting satellite can be heard at a particular location on the ground.

passband - The range of frequencies that will be allowed to pass through a filter without being diminished.

patch (*slang*) – 1. Autopatch. A device that interfaces an Amateur Radio repeater station to the telephone system. This allows a Ham using the repeater to make telephone calls over the air to any telephone. These devices have generally been replaced by cellular telephones. See: **autopatch**

2. (*antiquated term*) A phone patch. See: **phone patch**

PCB – Printed circuit board. See: **printed circuit board**

peak envelope power - The average power sent to the transmission line by the transmitter. Can be calculated by multiplying peak envelope voltage by 0.707. Abbreviated PEP. See: **PEP, PEV**

peak envelope voltage - The maximum voltage on a transmission line while a station is transmitting. Abbreviated as PEV.

peak-reading power meter – A measuring instrument that determines the peak power being run by a transmitter. The result is typically in watts. On some intermittent modes, such as single sideband, the usual standard average-power meter is not capable of measuring the actual output power. This requires a peak-reading power meter. See: **average power, average power meter**

peanut whistle (*slang*) - A modest Amateur Radio station setup with inexpensive equipment and limited antennas. See: **big gun, little pistol**

pecuniary interest – A reasonable likelihood or expectation that a person will receive cash or something else of value in exchange for a service rendered. Amateur Radio operators are forbidden from having any pecuniary interest in anything he or she does in regard to the hobby. In other words, a Ham may not be remunerated for such activities as relaying a message, reporting an accident, or the like. See: **Amateur**

pedestrian mobile (*slang*) - An Amateur Radio station operating while walking or running.

PEP - Peak envelope power. See: **peak envelope power**

perigee – A point in the orbit of a satellite in which it passes closest to the earth. See: **apogee**

personal (*slang*) – An operator's first name or preferred on-air name. This term originated as CB jargon but is usually frowned upon by Hams. Most prefer simply "name" or "handle."

PEV - Peak envelope voltage. See: **peak envelope voltage**

phone – An emission mode that uses voice modulation such as AM, SSB or FM.

phone patch (*antiquated term*) (*noun*) 1. A circuit or device that interfaces a telephone with an Amateur Radio transmitter and/or receiver. Such a device could allow a distant Ham to speak to a non-ham on a regular telephone through the interface with a Ham station of someone in the same area as the recipient of the call, avoiding long-distance charges. Phone patches were once common from military bases, persons in remote areas, or ships at sea to family members but are rarely needed now. See: **patch**

2. (*verb*) The process of conducting a conversation using Ham Radio interfaced to a regular telephone. Example: "I used to phone patch a missionary in South America through to his family here in town."

phonetic alphabet – A list of words that allows an operator on phone modes to spell out words to distinguish letters for clearer understanding. The list below is the NATO and International Telecommunications Union recommended version because it is more easily understood by native-speakers in many languages around the world. It is also the one recommended by the ARRL in the USA. That phonetic alphabet is:

A	Alpha
B	Bravo
C	Charlie
D	Delta
E	Echo
F	Foxtrot
G	Golf
H	Hotel
I	India
J	Juliet
K	Kilo
L	Lima (LEE – muh)
M	Mike
N	November
O	Oscar
P	Papa (Puh – PAH)
Q	Quebec (Kee – BECK)
R	Romeo
S	Sierra
T	Tango
U	Uniform

V	Victor
W	Whiskey
X	X-ray
Y	Yankee
Z	Zulu

picket fencing (*slang*) - A rapidly fluctuating choppy or fluttery signal, usually from a mobile station in motion. The name comes from the sound made by running a stick along a picket fence.

PICON – An acronym meaning "public interest, convenience and necessity." By law and as a trade-off for the creation of the Amateur Radio service, Hams should contribute to public service, assisting their fellow citizens. See: **public service**

pileup, pile-up (*slang*) – A large number of stations all calling a single station at the same time. A pileup usually occurs when a rare DX station is taking calls. See: **DX**

pink ticket (*slang*) – A notice of apparent violation, usually from the Federal Communications Commission, but sometimes also refers to a friendly notice from an Official Observer. See: **Official Observer**

pirate (*slang*) – 1. A station operating illegally by using on the air an existing or random call sign not assigned to him or her.
2. Any station transmitting illegally without possessing a properly issued license.

PKG (*CW abbrev.*) – "Package." "Packing." "Picking."

PL, PL tone - Private Line, a term trademarked by Motorola. Low frequency audio tones on a transmitted signal used to alert or control receiving stations. Same as CTCSS. See: **access code, Continuous Tone Coded Squelch System, CTCSS, tone**

PL-259 – A threaded male connector for coax cable. It is the most commonly used connector for Amateur Radio use and mates with a SO-239 connector. Sometimes referred to as a "UHF connector" though it is not recommended for use at UHF frequencies where the N-connector is a superior choice. See: **N-connector, SO-239, UHF connector**

plug - A male electrical connector designed to be inserted into a jack. See: **jack**

PM (*CW abbrev.*) – "Afternoon," "Evening."

portable – 1. An Amateur Radio station that is designed to be easily moved from place to place but is typically operated from a fixed position, not moving, as in a mobile station. See: **base station, fixed station, mobile**
2. (*slang*) Any operation of a station away from its officially licensed location.

portable designator - Additional identifying information that is added to a call sign to give listeners more information about the station's location. Such a designator typically follows the backslash sign on CW. This is often spoken as "slash" or "portable" on phone modes. Example: "This is N4KC slash maritime mobile." See: **maritime mobile, portable, slash**

positive offset - A repeater station's input (receiving) frequency is higher than its output (transmitting) frequency. Example: A repeater that transmits at 147.14 megahertz but is listening for users at 147.74 megahertz is said to have a positive offset of 600 kilohertz. See: **negative offset**

pot (*slang*) – Potentiometer. An electrical component that is a continuously variable resistor, often used for such purposes as a volume control.

power – The rate at which energy is consumed, expressed in watts. Power in a circuit is calculated by multiplying current by voltage.

Power Poles – See: **Anderson Power Poles**

power supply – In Amateur Radio today this generally means a device that converts 110 volts AC as provided by a normal wall socket in the USA into approximately 13.8 volts DC, the power requirement for most modern radio transceivers. The supply should also be able to handle the current requirements of the radio(s) to which it provides power. See: **linear power supply, switching power supply**

power up – 1. (*adjective*) The condition of a circuit when power has been applied so that it can operate.
2. (*verb*) To switch on the power to a circuit or equipment.

PRB – Private Radio Bureau. See: **Private Radio Bureau**

preamp, pre-amp – Preampflier. See: **preampflier**

preamplifier - A circuit in a receiver to boost weak signals before they have been otherwise modified by the receiver. The circuit also increases background noise, but can prove useful if that noise is minimal, as on higher HF and VHF/UHF frequencies. Often shortened to "preamp."

precedence – The priority of a message or traffic communicated via Amateur Radio. This might include 'informal," "formal," "health and welfare," "priority," or "emergency."

prefix – When speaking of an Amateur Radio call sign, the first part of the call including the number. Example: In the call sign WA7XYZ, the prefix is "WA7." See: **call district, call letters, call sign, suffix**

primary service – In cases where two or more radio services are required to share spectrum, the primary service is the one that always has precedence over a secondary service.

printed circuit board - A flat, non-conductive board on which electronic components are mounted and then connected by conductive traces etched ("printed") onto the material. Abbreviated as PCB.

priority channel – On a scanner or a radio with scanning capability, this is a channel that will be automatically selected if a signal is detected there regardless of where the function is in the scan order at the time.

priority traffic – A message or communication via Amateur Radio that is of considerable importance but not of an emergency nature. See: **emergency traffic, informal traffic, net, precedence, traffic**

Private Radio Bureau – The division of the Federal Communications Commission that directly administers Amateur Radio. Abbreviated PRB.

privileges - The frequencies, power levels, modes of communication, and other specific operating parameters that are permitted under the communications rules of a country.

PROB (*CW abbrev.*) – "Problem."

propagation - The means or path by which a radio signal travels from a transmitting station to a receiving station.

prosign – When using Morse code, these are one or two letters sent as a single character to abbreviate longer statements. Examples: "K" for, "Go ahead." "KN" for end of message but no breakers, please." "AR" for end of message. Two-letter prosigns should be sent as a single character. In this dictionary, all prosigns are simply designated as CW abbreviations.

PSE (*CW abbrev.*) – "Please."

PSK31 - Phase shift keying, 31 Baud. A popular radioteletype mode using tones generated from a computer soundcard. The mode is used primarily by amateur radio operators to conduct real-time keyboard-to-keyboard conversations. A variation is PSK63. See: **digital modes**

PTT - Push to talk or press to transmit. Activating a microphone and transmitter by pushing a button somewhere on the microphone, its base, or somewhere else nearby. When the transmission is complete, simply release the button to stop transmitting.

public service – Activities performed by Amateur Radio operators to benefit their communities and fellow citizens. According to law, Hams are to "serve the public interest, convenience and necessity," a phrase reduced to the meme PICON. See: **PICON**

pull the plug, pull the big switch (*slang*) - To shut down the station.

PWR (CW abbrev.) – "Power."

Q

Quebec ("Kee – BECK")

Dah – dah – di – dah

Q – 1. Quality factor. The response of a resonant circuit over a specific bandwidth. Generally a circuit with higher Q is more efficient but has a narrower bandwidth.

2. (*slang*) – A contact or QSO. Example: "Thanks for the Q and see you again soon." See: **QSO**

QCWA - Quarter Century Wireless Association. An organization for Amateur Radio operators who have been licensed for 25 years or more. Visit: **http://www.qcwa.org/**

QEX Magazine - A publication of the American Radio Relay League Published six times a year, the magazine is designed for more technically oriented Hams. Visit: **http://www.arrl.org/qex**

QRL – 1. Q-signal meaning, "Are you busy?" or "I am busy."

2. (*slang*) "Is the frequency in use?" Or, "The frequency is in use." Typically only used on CW or digital modes.

QRM – 1. Q-signal meaning, "Is my transmission being interfered with?" Or, "My transmission is being interfered with." Often used on voice and digital modes as well as CW.

2. (*slang*) (*noun*) Interference. Example: "There is too much QRM on this frequency for us to be able to carry on."

3. (*slang*) (*verb*) To cause interference. Example: "Let's move. We don't want to QRM the net."

QRN – 1. Q-signal meaning, "Are you troubled by static?" Or, "I am troubled by static." Often used on voice and digital modes as well as CW.

2. (*noun*) Static. Example: "The QRN is especially bad today."

QRO – 1. Q-signal for, "Shall I increase transmitter power?" Or, "I will increase power."

2. (*slang*) High power. Example: "You can tell he is running QRO."

QRP – 1. Q-signal meaning, "Shall I decrease transmitter power?" or "Decrease transmitter power." Often used on voice and digital modes as well as CW.

2. Using very low power while operating, typically five watts or less. Example: "I'm running QRP here today, just a watt output."

3. (*adjective*) Purposely running low-power. Example: "I'm going to compete in the QRP contest this weekend." Or, "They just released their new QRP rig."

QRP ARCI – QRP Amateur Radio Club International. An organization of Amateur Radio operators worldwide who are interested in low-power communication. The group produces a magazine, *QRP Quarterly*), organizes the annual "Four Days in May" conference each year in Dayton, Ohio, and sponsors various QRP contests and awards. Visit: **http://www.qrparci.org/** See: **Dayton, Four Days in May**

QRPp (*slang*) – Operating a Ham station while transmitting very low power, typically at one watt or less output. See: **QRP**

QRQ – 1. Q-signal for, "Shall I send (CW) faster?" Or, "Send faster."

2. (*slang*) Operating Morse code at very high speeds. Example: "He is a really good QRQ op."

QRS – 1. Q-signal for, "Shall I send (CW) slower?" Or, "Send slower."

2. (*slang*) – To send CW very slowly. Example: "Most Hams will happily QRS for newcomers if asked."

QRSS (slang) – Operating Morse code (CW) at an extremely low rate of speed, typically at less than one five-letter character per minute. Most often used in marginal propagation conditions and especially VLF and MF. See: **VLF, MF**

QRT – 1. Q-signal meaning, "Shall I stop sending?" Or, "I will stop sending." Often used on voice and digital modes as well as CW.

2. (*slang*) An operator is shutting down his station. Example: "I'm going to have to QRT now."

3. (*slang*) A strong suggestion to another operator to stop transmitting. Example: "Please QRT. You are interfering with emergency traffic."

QRU – 1. Q-signal meaning, "Have you anything for me?" Or, "I have nothing for you."

2. (*slang*) An indication an operator has no more to say and desires to end the contact.

QRV – Q-signal meaning, "Are you ready?" Or, "I am ready." Typically means the receiving station is ready to copy a CW message.

QRX – 1. Q-signal meaning, "When will you call again?" Or, "I will call again in ___ minutes." Often used on voice and digital modes as well.

2. (*slang*) The operator will be standing by for a short time but will return to the air. Example: On CW "QRX 5" means the operator will be standing by for five minutes. Or, "I have a phone call. QRX a minute."

QRZ – Q-signal meaning, "Who is calling me?" Often used on phone and digital modes as well where it is basically the equivalent to "CQ." Usually spoken as, "Q R Zed." See: **zed**

QRZ.com, "QRZ dot com, "QRZed dot com" – A popular Amateur Radio web site featuring a "flea market" area, news updates, and access to the FCC Amateur Radio database for lookup of Ham stations by call sign. Users are also encouraged to upload bio info and photos to the page where their FCC database results are displayed. Often shortened to "QRZ" or "QRZed." Example: "I just looked you up on QRZed and really like your shack." Visit: **www.qrz.com**

QSB – 1. Q-signal meaning, "Are my signals fading?" Or, "Your signals are fading." Often used on voice and digital modes as well as CW.

2. (*slang*) (*noun*) A signal that is fading, usually in and out, or conditions in general in which signals are tending to fade in and out. Example: "It is hard to copy you this morning with all the QSB."

Q-signals - A set of three-letter codes which are used by Amateurs as abbreviations or shortcuts. Though originally created for use in Morse code conversations, many of them are now commonly being used in digital and voice modes as well in the form of jargon, even if plain language might serve just as well or better. The more commonly used ones are included alphabetically in this section of the dictionary with an indication of whether or not it is used on all modes. When a Q-signal is followed by a question mark in CW, the meaning is in the form of a question. For a complete printable list of Q-signals typically used in Amateur Radio, visit: **http://www.arrl.org/get-on-the-air**

QSK – 1. Q-signal meaning, "Can you work break-in?" Or, "I can work break-in."

2. (*slang*) (*CW abbrev.*) (*noun*) – The capability of employing circuitry when using Morse code (CW) to be able to receive signals between dots and dashes while transmitting. This allows the receiving station to interrupt the communication without waiting for the transmitting station to finish, or for the sending operator to monitor what is happening on the frequency even while transmitting. See: **full breakin**

QSL – 1. Q-signal meaning, "Can you acknowledge receipt?" Or, "I can acknowledge receipt." Often used on voice and digital modes as well as CW.

2. (*slang*) "I understand," or "I copy."

3. (*noun*) Confirmation of a contact. This can be accomplished by sending a card or other written confirmation by regular mail or via several web-based eQSL services. See: **eQSL, LOTW, Logbook of the World**

4. (*noun*) A postcard-like confirmation of a Ham Radio contact, containing details of the contact. Can also be any other form of written confirmation such as a letter.

5. (*verb*) The act of confirming a Ham Radio contact, either by mail or electronically. Example: "Please QSL!" "Yes, I will be happy to QSL."

QSL bureau - Volunteer groups who help stations internationally to exchange QSL cards. They are typically maintained by a country's primary Amateur Radio organization, such as the American Radio Relay League in the USA. Operators are urged to keep the bureau stocked with self-addressed, stamped envelopes. When a DX operator works a number of stations, he or she will send a batch of cards to the bureau where they are sorted and, when a number of cards have been gathered, sent on to the stations who have envelopes on file. For more on the ARRL's QSL bureau service for incoming cards, visit: **http://www.arrl.org/incoming-qsl-service**. See: **bureau, BURO, QSL, QSL card, outgoing QSL bureau**

QSL card - A postcard-like confirmation of a Ham Radio contact. Also a confirmation by shortwave broadcast or commercial broadcast stations for reports by a listener of their signals having been received.

QSL manager - A volunteer who manages the receiving and sending of QSL cards for another Amateur Radio station. Typically the managed station, such as an operator in a rare country, makes so many contacts that he or she would have trouble managing or paying the postage for the volume of incoming QSL card requests.

QSO – 1. Q-signal meaning, "Can you communicate with _____ direct?" or "I can communicate with ____ direct." Often used on voice and digital modes as well as CW.
2. (*slang*) On-air, two way conversation. Example: "Thank you for the nice QSO." Sometimes pronounced "KEW –so."

QSO party (*slang*) – A contest in which stations from a particular state or geographic region attempt to contact as many other operators within or outside their state or region as they can during a set period of time. QSO parties traditionally are not nearly as intense as some other radiosport events. Visit: **http://www.hornucopia.com/contestcal/** See: **contest, radiosport**

QST – 1. Q-signal meaning, "Calling all radio amateurs." Typically used at the beginning of a broadcast from W1AW or when an Official Bulletin is about to be relayed by an OBS. See: **Official Bulletin**, **W1AW**

2. *QST*. The official magazine published by the American Radio Relay League. Available to members in both print and digital versions. Visit: **http://www.arrl.org/qst**

QSY – 1. Q-signal meaning, "Shall I change frequency?" or "I am changing frequency."

2. (*slang*) To change frequency. Often used on voice and digital modes. Example: "Let's QSY to 40 meters and see if the signals are better there."

QTH – 1. Q-signal meaning, "What is your location?" Or, "My location is _____." Often used on voice and digital modes as well as CW.

2. (*slang*) (*noun*) Current location. Example: "My QTH right now is Broad Street in front of the grocery store."

3. (*slang*) (*noun*) City in which the operator resides or the location of the operator's station. Example: "The QTH here is Springfield, Illinois." Though it is unnecessarily wordy, some operators say, "The home QTH is Springfield, Illinois." Or, "I'll sign now. I just got to my home QTH."

quad - A directional wire antenna consisting of two or more one-wavelength loops placed a quarter-wavelength apart. Quad loops can be constructed in a variety of shapes but are usually square.

quagi - An antenna, primarily for VHF and UHF, that employs some of the characteristics and construction of a quad and a Yagi. See: **quad**, **Yagi**

quarter-wave antenna – An antenna that has a length that is one-quarter-wavelength on the frequency for which it is intended to be used. To work properly, such an antenna requires a counterpoise such as radials, a quarter-wavelength piece of wire, or an automobile body. See: **counterpoise**, **radials**

quartz crystal - A piezo-electric mineral that can be inserted in an oscillator circuit and cut so it will vibrate at a particular frequency when an electric current passes through it. See: **crystal**, **crystal oscillator**

question pool - The set of approved questions that are used to put together Amateur Radio license examinations. Each class of license has a separate set of potential questions, and the pool is updated periodically.

Quicksilver Radio – A vendor that sells Amateur Radio equipment and supplies. Visit: **http://www.qsradio.com/**

R

Romeo ("ROH – me – oh")

Di – dah – dit

R – 1) Symbol for the resistance to the flow of electrical current in a circuit, measured in ohms.

2. (*CW abrev.*) "Received as transmitted," "Are."

RAC - Radio Amateurs of Canada.   The national Amateur Radio organization for Canadian Hams. Visit: **http://wp.rac.ca/**

RACES - Radio Amateur Civil Emergency Service. A service created and administered by the Federal Emergency Management Agency (FEMA) and the FCC, made up of licensed Hams that are certified by a civil defense agency and are able to communicate on Amateur Radio frequencies during drills, exercises and emergencies. Visit: **http://www.usraces.org/**

RadCom – Official magazine of the Radio Society of Great Britain. Visit: **http://rsgb.org/main/publications-archives/radcom/** See: **RSGB**

radials - Horizontal antenna elements designed to provide an electrical counterpoise to a vertically polarized antenna. Radials may be above ground, on the ground, or buried beneath the ground.

radiation resistance – The quantity of total resistance in an antenna system that causes RF energy to be radiated. This is typically the desired kind of radiation we want. The other kind of resistance in an antenna system is energy that is lost, usually in the form of heat. See: **resonance**

Radio Amateur Satellite Corporation - An educational organization. Its goal is to foster Amateur Radio's participation in space research and communication. The group has been responsible for designing, building and placing into orbit many Amateur Radio satellites. Abbreviated as AMSAT. Visit: **www.amsat.org**

radio check (*slang*) – An on-air query from an operator looking for an evaluation of his or her station's signal strength and audio quality.

Radio City – A vendor of Amateur Radio equipment and supplies. Visit: **http://www.radioinc.com/**

radio direction finding - Using directional antennas and receivers with calibrated reception metering to locate transmitting stations. Abbreviated as RDF. See: **bunny hunt, fox hunt, radio orienteering**

radio frequency - An alternating current which, if it is input to an antenna, generates an electromagnetic field suitable for such uses as wireless broadcasting. The frequencies of this type of alternating current are generally said to extend from 9 kilohertz to several thousand gigahertz.

radio frequency exposure – The amount, magnitude, and length of time of radio-frequency energy to which any individual is exposed. The Federal Communications Commission has established maximum values that are permitted and Amateur Radio operators must insure their stations meet or exceed those limits, not only for themselves but other persons who are near transmitting equipment and antennas.

radio frequency feedback - Distortion on a voice-modulated signal caused by radio frequency energy finding its way back into the microphone, microphone cable, connector or audio circuit of the transmitter.

radio frequency interference - Interference or spurious noise generated by a source emitting radio-frequency waves. Abbreviated as RFI.

radiogram – A formal piece of message traffic, usually one that is forwarded through traffic nets in the National Traffic System. See: **National Traffic System**

radio horizon - The most distance that radio signals can reliably travel using line-of-sight propagation. See: **line of sight**

Radio Mart – A vendor of Amateur Radio equipment and supplies. Visit: **http://www.radio-mart.net/**

radio orienteering – Using radio direction-finding techniques to determine the location of the listener or transmitting stations. See: **radio direction-finding**

radio shack (*slang*) – The area in which a Ham has set up his radio station. It can be a separate building but is usually just a corner in a room, office, garage, or basement area.

Radio Shack – A chain of stores that sell electronic parts and accessories.

Radio Society of Great Britain - The national organization for Amateur Radio operators in Great Britain. Visit: **http://rsgb.org/**

radiosport (*slang*) – On-air contesting. Though there is a wide variety of contests and competiveness, this typically refers to events in which Amateur Radio operators attempt to contact as many stations as they can within a specified time period in competition with other Hams. The more popular events include the CQ and ARRL DX contests, CQWPX, and Sweepstakes. Visit: **http://www.hornucopia.com/contestcal/** See: **contest, Sweepstakes, WPX**

radiotelephone – Another word for voice-modulated modes of radio communication.

radioteletype – A keyboard/printer mode of communications by radio. At one time, operators used surplus mechanical teletype terminals to enjoy this mode. Today most activity uses computer-generated audio tones or frequency-shift keying and display received text on computer monitors. See: **FSK, frequency shift keying, RTTY**

ragchewing (*slang*) - Chatting informally by way of radio. A "ragchew" is an informal conversation over the air.

random-wire antenna – An antenna consisting of any convenient length of wire, typically connected directly to a transmitter or antenna matching device. Performance of such an antenna can vary greatly.

Raspberry Pi – A very small basic computer that plugs into a TV and a keyboard, yet is capable of many of the things that a desktop PC can do. Typically, it uses the Unix operating system. It can be programmed for use in electronics projects as well as other applications and has found many uses in Amateur Radio. Visit: **https://www.raspberrypi.org/**

RBN – Reverse beacon network. See: **reverse beacon network**

R/C - Radio-controlled. Usually refers to model cars, boats or airplanes that are controlled using radio frequencies.

RCA plug – A type of connector commonly used to carry audio or video signals but often appear as key jacks, auxiliary connectors, or for other uses on Amateur Radio transceivers and equipment. Sometimes called phono or cinch plugs, named for Radio Corporation of America, an early developer. Visit: **https://en.wikipedia.org/wiki/RCA_connector**

RCV (*CW abbrev.*) – "Receive."

RCVR (*CW abbrev.*) – "Receiver."

RDF - Radio direction finding. See: **radio direction finding**

reactance - the opposition to current flow without dissipation of energy such as that of a capacitor or inductor in an AC circuit. Represented by the symbol "X."

reading the mail (*slang*) - Listening to an on-going on-the-air conversation without participating

receiver - A device or circuit that intercepts radio waves and converts them into signals that allow any intelligence carried on those waves to be heard and understood.

receiver incremental tuning - A control on a transceiver that allows the operator to vary the receive frequency a few kilohertz either side of the VFO frequency without affecting the transmitter frequency. Sometimes known as a clarifier. Abbreviated as RIT. See: **clarifier, transmitter incremental tuning**

reciprocal licensing, reciprocal operating authority – Allowing Amateur Radio licensees from foreign countries to operate in the United States without having to acquire a license in the USA. Such permission is based on agreements between the USA and governments of other nations and does not include all countries. Visit: **http://www.arrl.org/cept**) See: **CEPT agreement**

reflected power – Radio frequency power that is reflected back from the feed point of an antenna when there is an impedance mismatch. Non-radiated power may be dissipated as heat when the transmitter is mismatched to the antenna or load, or it may all be eventually radiated, depending on the type of feedline in use. See: **antenna analyzer, forward power, standing wave ratio, SWR**

reflector – The antenna element behind the driven element in a Yagi directional beam. See: **beam, Yagi**

refract – To bend. In relation to radio propagation, this is what the different layers of the ionosphere sometimes do with signals, bending them so they return to Earth at some distant point from their origination.

regulation – How well a power supply controls its voltage output.

relay – 1. A switch that is opened and closed by applying current to an electromagnet.

2. To pass along information from one station to another that is not hearing the first one well enough to copy his call sign and/or message.

3. To assist a net control station by conveying that a station he or she cannot hear is attempting to check into a net. See: **net, net control station**

4. To pass along a piece of formal radio traffic from one station to another. This is how the American Radio Relay League got its name.

remote control – QA situation in which the operator of a station is controlling it from some other location than where the transmitting equipment and antenna are situated. This can be from another room in the home or from the other side of the planet. Such activity has become more popular with the Internet, Skype and other aids. Many Hams now use remote control because of restrictive zoning or covenant limitations on antenna erection.

repeater, repeater station – Radio system, primarily for VHF and UHF frequencies, that receives incoming signals on one frequency and re-transmits them on another. Such stations are typically located at higher elevations and use good receiving antennas since they are intended to extend the range of communications for users' stations. Amateur Radio repeater stations are often sponsored by clubs or organized groups although many individual operators build and maintain them as well.

repeater directory – A listing of repeater stations around the country including their locations, input and output frequencies, open/closed status, and access tones, if any. Such a directory is published by the ARRL. There are also many on-line listings and directory smartphone apps available.

re-set the repeater, re-set the timer - Allowing the time-out circuit for a repeater station to return to zero and once again start timing a new transmission. On most repeaters, this occurs when a station on the input frequency stops transmitting. See: **timer**

resistor – An electrical component designed to offer opposition to the flow of current in a circuit. Its value is measured in ohms.

resonance – The point in a circuit, and especially antenna systems, at which inductive reactance and capacitive reactance are equal and cancel each other out leaving mostly radiation resistance. An antenna does not have to be resonant to perform well but an antenna system—everything from the output of the transmitter to the antenna—should be near resonance in order for the maximum transfer of power to take place. This is an often misunderstood aspect of antenna theory. Visit: **http://www.donkeith.com/n4kc/article.php?p=32**

resonant antenna – An antenna system in which inductive reactance and capacitive reactance are equal and only radiation resistance remains. See: **radiation resistance, resonance**

resonate – (*verb*) To make adjustments in an antenna to achieve a balance of inductive and capacitive reactance so that it is in a state of resonance. What an antenna system does when it is in resonance.

reverse beacon network - A network of mostly automated Amateur Radio stations listening to the bands and reporting stations whose Morse code signals are heard, when and how well. The stations use software called CW Skimmer to detect and report signals heard. Abbreviated as RBN. Visit: **http://www.reversebeacon.net/** See: **CW Skimmer**

RF - Radio frequency. See: **radio frequency**

RF burn - A burn caused by coming into contact with an RF voltage.

RF exposure – Radio frequency energy exposure. See: **radio frequency exposure**

RF feedback – Radio frequency feedback. See: **radio frequency feedback**

RF gain – Usually refers to a control on a receiver that allows the operator to increase or decrease the amount of amplification that is applied to an incoming signal. See: **AF gain**

RF ground – connection of Amateur equipment to Earth ground to eliminate hazards from RF exposure and reduce RFI. See: **RFI**

RFI - Radio frequency interference. See: **radio frequency interference**

RG-6, **RG-8**, **RG-8X**, **RG-58**, **RG-59** - Common types of coax cable used as feedlines in antenna systems. See: **LMR-200, LMR-400, 9913**

rice box (*slang*) – Amateur Radio equipment manufactured in China, Japan or elsewhere in the Orient

rig (*slang*) – Equipment used for transmitting and receiving in an Amateur Radio station.

RIG (*CW abbrev.*) (*slang*) Equipment used for transmitting and receiving in an Amateur Radio station.

ripple – A small amount of alternating current remaining after a DC power supply has supposedly rectified and filtered away all AC.

RIT - Receiver incremental tuning. See: **receiver incremental tuning**

R&L Electronics – A vendor of Amateur Radio equipment and supplies. Visit: **http://www.randl.com/shop/catalog/**

rock (*slang*) – Piezo-electric crystal that determines the frequency of a transmitter.

rockbound (*slang*) – Operating with a transmitter that's frequency is controlled by a crystal and thus restricted to one frequency.

roger – "I understand."   "I copied your full transmission." Sometimes varied to indicate not full copy, such as, "Roger on most but missed your name."

roger beep (*slang*) - A tone or sound that is heard, usually in Amateur Radio on a repeater station, when a radio operator un-keys his or her microphone.   Lets other users know the station is no longer transmitting. Originated in the Citizens Band service where individual stations employed such a tone on their own transmissions. Such practice is not encouraged in Amateur Radio. See: **un-key**

roofing filter - A type of filter found in many modern receivers, typically in the circuit just after the first receiver mixer. Its purpose is to reject stronger signals on adjacent frequencies before more processing of the desired signals take place. See: **filter, mixer**

rotator – An electric motor-driven device attached to an antenna mast to turn so the antenna can be pointed in the desired direction. Many years ago, this device was called a "rotor" and sometimes is today.

rotor (*slang*) (*antiquated term*) – Former word often used to describe a device used to mechanically turn a beam antenna to a selected direction. The proper name is "rotator." See: **rotator**

roundtable (*slang*) – A group of Amateur Radio stations in a conversation, taking turns transmitting. See: **net**

rover (*slang*) – An Amateur Radio mobile station that drives from one spot to another, operating from multiple grid squares, counties, or states over a period of time. Rovers are typically active during county-hunter nets and in contests. See: **county hunter, grid square, radiosport**

RPT (*CW abbrev.*) – "Repeat," or "Report (signal)."

RPRT (*CW abbrev.*) – "Report."

RPTR (*CW abbrev.*) – "Repeater."

RSGB – Radio Society of Great Britain. The national organization for Amateur Radio operators in Great Britain. Visit: **http://rsgb.org/**

RSQ – Readability, signal strength, and quality. A three-digit signal reporting system used to tell an operator how his digital mode signals are being received. Though many digital operators continue to use the well-accepted RST system, RSQ is finding more users because it is more applicable to those modes.

Note that even numbers are not used on the signal-strength and readability scales. The RSQ system is:

READABILITY

1 -- No copy, undecipherable

2 -- 20% copy, occasional words distinguishable

3 -- 40% copy, readable with difficulty, many words missed

4 -- 80% copy, readable with difficulty

5 -- 95% or more copy, perfectly readable

STRENGTH

1 -- Barely perceptible trace

2 -- Not used

3 -- Weak trace

4 -- Not used

5 -- Moderate trace

6 – Not used

7 -- Strong trace

8 -- Not used

9 --Very strong trace

QUALITY

1 -- Splatter over much of the spectrum

2 -- Not used

3 -- Multiple visible pairs

4 -- Not used

5 -- One easily visible pair

6 -- Not used

7 -- One barely visible pair

8 -- Not used

9 -- Clean signal - no visible unwanted sidebars

RST - Readability, signal strength, and tone. The three-digit signal reporting system used to tell an operator how his station is being received. For voice modes, only the first two numbers are used. Tone applies only to CW transmissions.

The RST system is:

READABILITY

1 -- Unreadable

2 -- Barely readable, occasional words distinguishable

3 -- Readable with considerable difficulty

4 -- Readable with practically no difficulty

5 -- Perfectly readable

SIGNAL STRENGTH

1 -- Faint signals, barely perceptible

2 -- Very weak signals

3 -- Weak signals

4 -- Fair signals

5 -- Fairly good signals

6 -- Good signals

7 -- Moderately strong signals

8 -- Strong signals

9 -- Extremely strong signals

TONE

1 -- Sixty cycle AC, or less, very rough and broad

2 -- Very rough AC, very harsh and broad

3 -- Rough AC tone, rectified but not filtered

4 -- Rough note, some trace of filtering

5 -- Filtered rectified AC but strongly ripple-modulated

6 -- Filtered tone, definite trace of ripple modulation

7 -- Near pure tone, trace of ripple modulation

8 -- Near perfect tone, slight trace of modulation

9 -- Perfect tone, no trace of ripple or modulation of any kind

RTTY - Abbreviation for radioteletype. See: **radioteletype**

rubber duck (*slang*) - A shortened flexible antenna usually used with hand-held scanners and transceivers. Not a particularly effective antenna but a trade-off for flexibility and ease of use.

RX (CW abbrev.) – "Receive," "Receiver."

RY – A letter combination sent to test the response of a radioteletype setup. Used because the code that represents these two characters requires the greatest variation in bits from one to the other.

S

Sierra ("See – AIR – uh")

Di – di – dit

SA (*CW abbrev.*) – "South America."

safety interlock -- A switch designed to automatically turn off electricity to a piece of equipment if its cover is removed. This protective device should never be intentionally defeated.

SASE - Self-addressed, stamped envelope. Recommended to be sent with a mailed request for a QSL card. See: **QSL card**

SATERN – Salvation Army Team Emergency Radio Network. A group affiliated with The Salvation Army dedicated to assisting the organization during times of emergency by providing communications when normal means are not available. Visit: **http://www.satern.org/**

scan – Programming a receiver to continually sample designated frequencies or a range of frequencies. When a signal is detected, the scan stops temporarily so the operator can halt the scanning process and listen to the incoming signal.

scanner – A radio receiver with the capability of being programmed to listen to only designated frequencies or a range of frequencies, then to stop temporarily when a station is heard.

schedule (*slang*) - A scheduled on-air conversation at regular time and frequency with another Ham. May be a one-time or a recurring event.

schematic – A drawing or diagram of an electronic circuit.

schematic symbol - A symbol used to represent a component on a schematic diagram. Visit: **https://www.edrawsoft.com/circuit-symbols.php**

School Club Roundup – An operating event sponsored by the American Radio Relay League in which school Amateur Radio clubs attempt to contact each other as well as make contacts with other Hams. Visit: **http://www.arrl.org/school-club-roundup**

screwdriver antenna – A vertical antenna that employs an electric motor to raise and lower the element to attempt to more easily find a match. Some designs actually use motors made for use in electric screwdrivers, thus the name. These antennas are most often used in mobile applications. See: **match, mobile**

SDR – Software defined radio. A radio transmitter and/or receiver in which operations typically done by hardware components (such as amplifiers, detectors, filters, etc.) are accomplished with software, either on an externally attached personal computer or an internal processor dedicated to the radio.

search - A receiver feature that allows the user to set up a frequency range for the receiver to scan. The receiver will then pause on a frequency if a signal is heard.

SEC - Section Emergency Coordinator. See: **Section Emergency Coordinator**

secondary allocation, secondary status – A situation in which a frequency band or portion of a band is shared with one or more other communication services. On a shared band, if the Amateur Radio service is the secondary user, Hams must not cause interference to the primary user.

section – A geographical division of the Field Organization of the American Radio Relay League. Each section has a Section Manager (SM) and other volunteer positions. Sections are often states but in areas with a higher Ham population, the areas are smaller. Example: Georgia and Alaska are sections. Pennsylvania is broken up into Eastern and Western. Visit: **http://www.arrl.org/sections** See: **ARRL, Section Manager, SM**

Section Emergency Coordinator - A volunteer in the ARRL Field Organization who organizes emergency response by Hams in his or her section. Abbreviated as SEC.

Section Manager - Section Manager. A volunteer in the ARRL Field Organization who organizes Amateur Radio activity in his or her section, appoints Hams to other volunteer positions, and represents the section to the rest of the ARRL's Field Organization.

SED (CW abbrev.) – "Said."

selectivity – The ability of a receiver to reject undesired signals adjacent to the desired signal.

self-spotting (*slang*) – An operator sending his or her own call sign to a DX cluster in an attempt to get other stations to come work the self-spotter. This is generally not encouraged. See: **DX cluster, spotting**

self-supporting – A tower or mast that is designed to be erected and to stay upright without the use of supports or guys. See: **guy**

semi-break-in - Employing circuitry when using Morse code (CW) to be able to receive signals between characters while transmitting. Semi break-in enables an operator to listen to other signals between individual characters and/or words. This allows the receiving station to interrupt the communication without waiting for the transmitting station to finish. This is sometimes referred to as "QSK," from the Q signal for "I can operate break-in." See: **break-in, full break-in, QSK, Q signal**

sensitivity – A measure of how well a receiver can pick up a weak signal.

separation – The frequency difference between a repeater station's input (receive) frequency and output (transmit) frequency. This is sometimes called "split." See: **split**

SET – Simulated Emergency Test. See: **Simulated Emergency Test**

SEZ (CW abbrev.) – "Says."

SFI - Solar flux index. See: **solar flux index**

shack (*slang*) – The area in which a Ham has set up his radio station. It can be a separate building but is usually just a corner in a room, garage, or basement area.

Sherwood rankings – A popular and exhaustive ranking of commercially available Ham Radio receivers and transceivers by Rob Sherwood NCØB. The list is ranked by third order dynamic range, narrow spaced, a criterion that may or may not be of most importance depending on individual user needs. Visit: **http://www.sherweng.com/table.html** See: **dynamic range**

short path - A signal path that is the shortest route from transmitting station to receiving station. The reciprocal path is called long path. See: **great circle route**, **long path**

short skip - Propagation of a radio signal by the ionosphere over a few hundred miles or less.

shortwave – The portion of the radio spectrum usually defined as HF or high frequencies, between 3 and 30 megahertz. See: **high frequency**

shortwave listener – A hobbyist who enjoys listening to the shortwave radio frequencies and international broadcasts. Abbreviated as SWL. See: **shortwave**

sidewinder (*slang*) (*antiquated term*) – An Amateur Radio operator who uses the single-sideband voice mode.

SIG (*CW abbrev.*) – "Signature," "Signal."

sign, sign out, signing off (*slang*) – To end a contact, leave the air and close down the station.

signal – Electrical or electromagnetic impulse that conveys information.

signal generator - A device or circuit that can generate a low-power signal at a designated frequency, typically used for testing purposes.

signal report – An evaluation by a Ham of a another station's readability, signal strength, and technical quality. See: **RSQ**, **RST, S-meter**

signal strength meter – A meter on a receiver designed to give the relative strength of signals. Usually called an S-meter. See: **S-meter**

SIGS (*CW abbrev.*) – "Signals."

silent key - A deceased Amateur Radio operator. Abbreviated as SK.

simplex - An operating mode in which the transmit and receive frequencies are both the same. See: **duplex**

Simulated Emergency Test - A nationwide emergency communications exercise held annually. Administered by the American Radio Relay League. Abbreviated as SET. See: **Amateur Radio Emergency Service**

single sideband - A voice emission mode in which the carrier is greatly reduced and one sideband of the signal is filtered out. See: **balanced modulator, SSB**

Six Meter International Radio Klub – Organization formed to promote the use of the 6-meter Amateur Radio band. Abbreviated as SMIRK. Visit: **http://www.smirk.org/**

SK (*CW abbrev.*) – 1.) "This is my last transmission." "End of contact." Sent as single character, di-di-di-dah-di-dah.
2. Abbreviation for "Silent key," a deceased Amateur Radio operator.

SKCC – Straight Key Century Club. See: **Straight Key Century Club**

SKED (CW abbrev.) – "Schedule." A scheduled on-air conversation at a regular time and frequency with another Ham. See: **schedule**

skimmer (*slang*) - A multi-channel Morse code (CW) decoder and analyzer program. The system is able to detect and decode most signals heard within a broad range of frequencies and display the call signs of the stations on a web site.

skip (*slang*) – The electromagnetic phenomenon in which signals are reflected or refracted by various layers in the Earth's atmosphere and come back down, sometimes very far away from the original point of origin. The area between those two points, in which the transmission cannot be heard, is called the skip zone. See: **propagation, skip zone**

skip zone (*slang*) – A geographic area in which shortwave signals cannot be heard because the transmitting station is too far away for ground wave propagation and too near for sky wave propagation. See: **ground wave, sky wave**

SKN – Straight Key Night. See: **Straight Key Night**

skyhook (*slang*) – A very large antenna.

SKYWARN – An organization of trained volunteer storm spotters who work with the National Weather Service. Many of its members use Amateur Radio to communicate their reports of dangerous weather. Visit: **http://skywarn.org/**

sky wave – The reflection of radio waves off the ionosphere.

skywire (*slang*) – A large horizontal loop antenna. See: **horizontal loop**

slash (*slang*) – The character: / or "front slash." In Ham Radio, it is typically used to append a portable designator to a call sign. Used on CW as well as phone modes. Example: "This is WA8QQQ slash KH6 in Honolulu."

S-line (*antiquated term*) - A commercially manufactured series of transmitters, receivers, amplifiers, and other Amateur Radio accessories made and sold by the Collins Radio Company from the 1950s to the 1970s.

slim (*slang*) – An unscrupulous person who pretends to be a DX station or other highly desired contact, taking calls and giving signal reports. See: **pirate**

slop bucket (*slang*) (*antiquated term*) – Derogatory term for single sideband, used in the early days of the development of the mode by those who did not like it.

sloper (*slang*) – A dipole or end-fed wire antenna with one end much higher in elevation than the other.

slow-scan television - A picture transmission method to transmit and receive static pictures via radio. This mode is mostly used by Amateur Radio operators on the HF frequency bands. See: **SSTV**

SM - Section Manager. A volunteer in the ARRL Field Organization who organizes Amateur Radio activity in his or her section, appoints Hams to volunteer positions, and represents the section to ARRL headquarters.

SMA – A type of small coaxial cable connector often used to attach an antenna on VHF/UHF portable transceivers.

SMC connector – Sub-miniature C-type. A type of radio-frequency connector typically used for antenna connections to a radio.

S-meter - Signal strength meter. A meter typically part of a receiver that indicates the relative strength of comparable signals. Calibrated in S-units and decibels. See: **decibel, dB, S-unit**

Smith chart - A graphical aid designed to assist in visualizing relationship in and solving problems with transmission lines and matching circuits.

SMIRK – Six Meter International Radio Klub. See: **Six Meter International Radio Klub**

SN (CW abbrev.) – "Soon."

S/N - Signal-to-noise.

S/N ratio – The ratio between a signal and other noise in the receiver. This is one factor to consider in evaluating the performance of a receiver.

SO-239 – A female threaded coax connector, often used on the output of Amateur Radio transceivers. It mates with the PL-259 male connector. See: **PL-259**

soapbox (*slang*) – The section in Ham magazine coverage of radiosport events in which participants talk about their own experiences.

software defined radio - A radio transmitter and/or receiver in which operations typically done by hardware components (such as amplifiers, detectors, filters, etc.) are accomplished with software, either on an externally attached personal computer or an internal processor dedicated to the radio. See: **SDR**

solar flux index - A measurement of solar particles and magnetic fields from our sun that reach our atmosphere, as reported by the Penticton Radio Observatory in British Columbia, Canada. It can vary from values below 50 to values in excess of 300. A higher solar flux index can mean better radio-frequency propagation on the high frequency bands. Abbreviated as **SFI**.

solder – (pronounced "SOD er") 1. (*noun*) A metal alloy that melts at comparatively low heat employed for joining wires or other metals electrically and physically.
2. (*verb*) To join together with solder. See: **soldering gun, soldering iron**

soldering gun, soldering iron – (pronounced "SOD er ing") A device used to heat and melt solder for the purpose of joining two wires or metals together electrically or mechanically. See: **solder**

SOS (*CW abbrev.*) – A call that indicates that a life-threatening event is being reported or relayed, a distress call typically used on Morse code or digital modes as opposed to "Mayday" on voice transmissions. See: **Mayday**

SOTA – Summits on the Air. An award program for Radio Amateurs and shortwave listeners that encourages portable operation in mountainous areas. Visit: **http://www.sota.org.uk/**

SP – 1. (*CW abbrev.*) – "Speaker," "Speed (of sending code)."
2. Abbreviation for "speaker."

spark gap (*antiquated term*) - an early type of transmitter design that employed electrical sparks to generate radio frequency signals.

speaker mic, speaker microphone - An accessory usually used with a hand-held transceiver that contains in one unit both a speaker and a microphone.

special event call sign - A special temporary call sign consisting of a single letter, a number, and another single letter, issued to special event stations or for other distinct purposes. Example: The call sign N9N was issued for the special event station commemorating the fiftieth anniversary of the journey to the North Pole by the submarine USS *Nautilus*. Visit: **http://www.1x1callsigns.org/**

special event station – An operation of an Amateur Radio station designed to acknowledge or commemorate an event, anniversary, festival, etc. Many Ham Radio clubs use such operations to gain public exposure for the hobby. Other Hams enjoy collecting QSL cards or certificates received for contacting such operations.

spectrum – As it pertains to radio, the range of electromagnetic frequencies that are typically used to propagate signals, generally considered to lie between 10 kilohertz and 300,000 megahertz.

Spectrum Monitor – An online magazine devoted to all aspects of hobby radio including Ham Radio. Delivered as an e-zine PDF. Visit: **http://www.thespectrummonitor.com/**

spectrum scope – A scope or screen display used to visually monitor a wide band of frequencies at the same time. Sometimes called a "pan adapter." See: **panadapter**

speech processor - A circuit that increases the average level of the audio before it modulates a transmitted signal. Improper adjustment can cause considerable interference to other stations. See: **compression**

splatter - a type of spurious emission that can cause interference to stations on nearby frequencies. Splatter occurs when a transmitter's carrier signal is modulated too heavily. See: **spurious emission**

split – 1. The frequency difference between a repeater station's input (receive) frequency and output (transmit) frequency. This is sometimes called "separation."

2. (*slang*) A DX station operating in such a manner that he is transmitting on one frequency but listening for callers on another nearby frequency. This is to keep a large number of calling stations from covering up the DX station.

sporadic-E – A type of radio signal propagation that occurs when random patches of intense ionization form in the E-layer of the ionosphere and refract higher frequency signals rather than absorb them. This phenomenon typically occurs on 10 meters or above and is especially prevalent on 6 meters and 2 meters in springtime and again around mid-December. See: **E-layer, E-skip**

spots (*slang*) – Listing of DX stations heard or worked on DX spotting web sites. See: **cluster, DX cluster, DX spotting, spotting**

spotting (*slang*) - A process in which stations report hearing ("spotting") or making contact with other stations to a web site where it may be seen by other operators. This allows operators who wish to talk to those stations to go to that frequency and attempt to make the contact. See: **cluster, DX cluster, DX spotting, spots**

sprint (*slang*) – An on-air contest of very limited duration, often as short as an hour or two. See: **contest, radiosport**

spur (*slang*) - Spurious signals. Undesired signals that can come from various sources. They are more serious when present in the output of a transmitter since they can cause interference to other stations. They can also occur in a receiver and make it difficult to hear certain signals.

spurious emission – An undesirable electromagnetic signal that occurs outside the necessary bandwidth of a transmission, such as harmonics, splatter, and the like. See: **harmonic, splatter**

SQL – Abbreviation of "squelch." See: **squelch**

squelch - A circuit that mutes the receiver when no signal or only marginal signals are present, thereby eliminating having to listen to band noise or unreadable signals. Abbreviated as SQL.

squelch tail (*slang*) - A very short bit of noise heard on a repeater station after the end of a radio transmission by a user and before the the receiver's squelch circuit has been reactivated.

SRI (CW abbrev.) – "Sorry."

SSB - Single sideband. A voice emission mode in which the carrier is greatly reduced and one sideband of the signal is filtered out.

SSN - Sunspot number. See: **sunspot number**

SSTV – Slow-scan television. See: **slow-scan television**

standing wave ratio - A ratio between forward and reflected power in an antenna system. It shows how efficiently radio-frequency power is transmitted from a power source (a transmitter or amplifier) through a transmission line (coax, open wire feed line, a single wire), into a load (an antenna). Abbreviated SWR. See: **forward power, reflected power, SWR**

SteppIR – A company that manufactures commercially made antennas for Radio Amateur use. Their primary product is a Yagi-type beam that uses small motors and copper strips to vary the lengths of the beam elements to bring them into resonance. See: **beam, Yagi**

stinger (*slang*) – A small attachment to a whip antenna to make it closer to resonance on a particular band. It typically would contain a coil and a short length of whip.

STN (*CW abbrev.*) – "Station."

straight key (*slang*) - a non-electronic switch mounted on a pedestal used for forming the dots and dashes of Morse code. It typically uses a single spring-loaded lever that is pressed down and released by the operator to close and open a circuit in order to form dots and dashes. See: **key, Morse code**

Straight Key Century Club - An organization of Amateurs who prefer using mechanical keying devices when operating CW. Abbreviated SKCC. Visit: **http://www.skccgroup.com/**

Straight Key Night – Annual operating even sponsored by the American Radio Relay League, held on New Year's Eve/New Year's Day. Operators use CW (Morse code) to hold informal conversations, typically using vintage or otherwise unique equipment and straight keys with which to send the code. Abbreviated as SKN. Visit: **http://www.arrl.org/straight-key-night** See: **ARRL, straight key**

strays – 1. (*slang*) (*antiquated term*) Static.
2. Short informational notices in the pages of *QST Magazine*. See: *QST Magazine*

strength unit – A unit of signal strength measurement used on S-meters. Abbreviated as S-unit. See: **S-meter**

stub - A length of transmission line that is used to help bring an antenna system into resonance. By choosing the proper length and the characteristic impedance, and having one end open or shorted, a stub becomes in effect a capacitor or inductor and can be used to achieve a match when inserted at a selected point in the regular transmission line. See: **feed line, match, resonance**

subaudible tone, sub-audible tone – A tone that cannot be easily heard on a radio signal because it is below the normal hearing range of humans. Such tones are used for a number of purposes, including controlling access to repeater stations.

suffix - When speaking of an Amateur Radio call sign, the part of the call that follows the number. Example: In the call sign WA7XYZ, the suffix is "XYZ." See: **call district, call letters, call sign, prefix**

S-unit - Markings on the scale of an S-meter, derived from a system of reporting signal strength from S1 to S9 that was developed as part of the RST code. See: **RST, S-meter**

superheterodyne – A type of radio receiver circuit in which an internal signal is generated to mix with a received signal to form a signal at a new frequency. The receiver then processes that new signal, which is now at a frequency that can be more efficiently handled.

SUM (CW abbrev.) – "Some."

Summits on the Air - An award program for Radio Amateurs and shortwave listeners that encourages portable operation in mountainous areas. Visit: **http://www.sota.org.uk/**

sunspot - A storm on the surface of the sun. Such activity can affect the ionization level of the ionosphere. Generally, more spots mean better long distance propagation. See: **sunspot cycle, sunspot number**

sunspot cycle - The regular periods of decline and incline in the numbers of sunspots on the surface of the sun. This is a relatively predictable eleven-year cycle.

sunspot number - An arbitrary numerical value that is used to describe how active the sun's surface is over a period of time. The presence of more sunspots is typically good for radio propagation on the HF portion of the spectrum. Sunspot minima and maxima run in 11-year cycles. Abbreviated as SSN.

superheterodyne - A type of receiver design in which an incoming signal is beat—or heterodyned—with an internally generated signal to create a signal at a frequency that can be more efficiently processed.

surface mount - A method for producing electronic circuits in which the components are mounted or placed directly onto the surface of printed circuit boards. See: **PCB, printed circuit board**

SW 1. (*CW abbrev.*) – "Switch."
2. Abbreviation for "shortwave."

Swan Electronics – A former manufacturer of Amateur Radio equipment.

Sweepstakes – A major radiosport event sponsored each November by the ARRL. Visit: **http://www.arrl.org/sweepstakes** See: **radiosport**

switching power supply - A power supply that uses switching transistors to convert AC to DC rather than a transformer, diodes and electrolytic capacitors like a linear supply. See: **linear power supply, power supply**

switcher (*slang*) – A switching power supply. See: **power supply, switching power supply**

SWL - Shortwave listener. See: **shortwave listener**

SWR - Standing wave ratio. See: **standing wave ratio**

SWR meter – A measuring device used to determine the standing wave ratio of an antenna system.

T

Tango

Dah

T (*CW abbrev.*) – Zero ("dah"). See: **cut numbers**

talk-in (*slang*) – Giving driving directions to incoming visitors to a Ham Radio event or venue, most often on a VHF or UHF repeater.

talk-in frequency (*slang*) – The designated frequencies or channels on which operators offering driving directions to a Ham Radio event or venue can be found.

talk out (*slang*) – To talk longer than the amount of time at which a repeater station's talk out timer is set and have the repeater stop transmitting. See: **courtesy beep, talk out timer**

talk out timer (*slang*) – A circuit in a repeater station that will turn off the repeater's carrier if a user talks longer than the amount of time for which it is set. This is to prevent someone from dominating the repeater and also to allow the transmitter's safe duty cycle to be maintained. See: **courtesy beep, duty cycle, talk out**

TAPR - Tucson Amateur Packet Radio Corp. An organization that supports research and development in Amateur Radio digital communications modes. Visit: **https://www.tapr.org/** See: **packet radio**

Technician, Technician class – The currently available introductory class of Amateur Radio license in the USA. See: **Amateur Extra, General**

telegraphy - The transmission of information in Morse code. Pronounced "tuh LEG ruh fee." See **CW, Morse code**

telemetry - A one-way transmission using radio that carries information that might be used for tracking and to carry measurement data.

telephony - The transmission of information in a voice mode. Pronounced "tuh LEF oh nee." See: **AM, FM, SSB**

telescoping antenna - An antenna that slides into and out of itself to become shorter or longer for storage or matching purposes.

temporary state of emergency – A declaration by the Federal Communications Commission that an emergency has occurred in a designated area. Specific rules then apply to all Amateur Radio operators in that area for the duration of the emergency.

Ten Tec – A major manufacturer of Amateur Radio equipment. Visit: **http://www.rkrdesignsllc.com/**

Ten Ten International - An organization of Amateur Radio operators dedicated to maintaining interest in operating on the 10-meter Ham band. Abbreviated as 10-10. Visit: **http://www.ten-ten.org/**

terminal node controller – A device used as a modem to encode and decode packets of digital data onto a signal for use in packet radio. Abbreviated as TNC. Visit: **https://www.tapr.org/** See: **modem, packet radio**

TEST (*slang*) (*CW abbrev.*) – On-air contest, radiosport event. The term may also be used on voice modes, too. Example: "CQ TEST CQ TEST DE PJ1XX."

test session – An event in which the Amateur Radio license examination is administered by Volunteer Examiners. See: **Volunteer Examiner**

Texas Towers – A vendor of Amateur Radio equipment and supplies. Visit: **http://www.texastowers.com/**

The League (*slang*) - The American Radio Relay League. See: **American Radio Relay League, ARRL**

third-party – An unlicensed person for whom traffic might be passed via Amateur Radio. See: **third party communications, third party traffic**

third-party communications, third-party traffic - Messages passed from one Amateur Radio operator to another on behalf of a third person. Amateurs may not receive any payment for handling such messages.

third-party communications agreement - An official understanding, usually confirmed by treaty, between one country and another allowing licensed Amateur Radio operators in both countries to pass along third-party communications from one to the other. See: **third-party communications**

third-party operation – Operation of an Amateur Radio station by an unlicensed person. Such operation must be under the direction of a control operator.

THRU (*CW abbrev.*) – "Through." "Threw."

ticket (*slang*) - An Amateur Radio license.

TIL (*CW abbrev.*) – "Until."

time out (*slang*) - To talk longer than the amount of time at which a repeater station's talk out timer is set and have the repeater stop transmitting. See: **courtesy beep, talk out timer**

timer (*slang*) - A circuit in a repeater station that is set to shut off the repeater's carrier if a user talks longer than the amount of time for which it is set. This prevents someone from dominating the repeater and allows the transmitter's duty cycle to be observed. See: **courtesy beep, talk out**

TKS (*CW abbrev.*) – "Thanks."

TMW (*CW abbrev.*) – "Tomorrow."

TMRW (*CW abbrev.*) – "Tomorrow."

TNC – Terminal node controller. See: **terminal node controller**

TNX (*CW abbrev.*) – "Thanks."

tone - 1. (*slang*) A sub-audible tone needed for computer station access See: **access code**
2. An audio-frequency signal generated at a pre-determined constant or varying frequency used to test modulation, signal waveform, power output, and other parameters.

tone access – The necessity for a sub-audible tone of a specific frequency that must be inserted into an operator's audio in order for a repeater station to re-broadcast the incoming signal. See: **access code**, **tone**

tone pad – A device with twelve or sixteen numbered touch keys each of which can generate a standard telephone multi-frequency two-tone dialing signal. Tone pads are typically built into hand microphones that are supplied with VHF/UHF transceivers since they are often used to perform various functions on a repeater station. See: **autopatch**

top band (*slang*) – the 160-meter Ham band, so called because it has always been (and is as of this writing) the highest assigned frequency band when considering wavelength.

toroid - A donut-shaped device usually made of ferrous materials that can be used as an inductor, such as on coax feedlines, to choke off unwanted RF energy that might travel the outside of the cable's shield.

T/R - Transmit/receive.

TR (*CW abbrev.*) – "Transmit," "Transmitter."

traffic (*slang*) - A message or messages sent by radio. Such communication can be formal or informal. See: **emergency traffic, informal traffic, formal traffic, National Traffic System, net, precedence, NTS**

transceiver – A communication device that is capable of both transmitting and receiving radio-frequency signals, usually employing much of the same circuitry to perform both processes.

transducer (*antiquated term*) – An early word for antenna.

transient - A very short burst of energy on a power line, usually lasting for fractions of a second. Transient spikes can still cause damage to equipment connected to the power line.

transmatch – Another word for "antenna tuner." See: **antenna tuner, matchbox**

transmission line – 1. Another term for feedline. See: **feedline**
2. A set of wires for carrying commercial power cross country.

transmitter - A circuit that generates radio-frequency signals of enough strength to allow for communication. Abbreviated as XMTR.

transverter - A circuit in or attached to a transceiver or transmitter that allows the radio to operate on other bands. This capability is typically employed so a transceiver designed only for HF can be used on VHF or UHF bands. Abbreviated as XVTR.

TRBL (*CW abbrev.*) – " Trouble."

tropo (*slang*) – Term for tropospheric ducting propagation. Pronounced "TROP oh." See: **tropospheric ducting**

tropospheric ducting – The propagation of radio signals via bending and ducting along weather fronts. Such phenomena occur in the lowest layer of the Earth's atmosphere, the troposphere, and is most present on frequencies above 30 megahertz. See: **propagation**

T/R switch – A switch or relay used to change an antenna feedline from the transmitter output to the receiver input.

trunked radio system - A computer-controlled scheme used in two-way radio communications that allows more efficient use of relatively few radio frequency channels while helping prevent eavesdropping. Abbreviated as TRS.

TRX (*CW abbrev.*) – "Transceiver," "Transmitting."

TT (*CW abbrev.*) – "That."

TU (*CW abbrev.*) – "Thank you."

tube – Vacuum tube. See: **vacuum tube**

tuner (*slang*) – Matchbox, antenna matching device, "antenna tuner."

TVI (*slang*) – Interference to a television set by an Amateur Radio station's transmissions. Also sometimes used in regard to interference to Amateur Radio receivers by a television set.

twin-lead – A type of balanced twin-conductor feedline encased in plastic that keeps the two wires an equal distance apart. Twin-lead was once used extensively for TV antenna feedline.

twisted pair (*slang*) (*antiquated term*) – The common telephone. The term is used to make it clear the operator is not talking about the phone mode of radio transmission. Example: "If we lose each other in the static, give me a call on the twisted pair." See: **landline**

two-tone test - A method of testing the audio or power output of a single sideband transmitter by feeding two audio tones of different frequencies into the microphone input of the transmitter and observing the output on an oscilloscope. The two simultaneous tones more closely represent the tonal characteristics of the human voice.

TX (*CW abbrev.*) – "Transmitter," "Transmit." See: **transmitter**

TXT (*CW abbrev.*) – "Text," typically referring to the part of a formal piece of radio traffic that contains the content of the message.

U

Uniform

Di – di – dah

U (*CW abbrev.*) – "You."

UHF – Ultra high frequency. See: **ultra high frequency**

UHF connector – See: **PL-259**

ULS – Universal Licensing System. See: **Universal Licensing System**

ultra high frequency - The portion of the radio-frequency spectrum between 300 and 3000 megahertz. Abbreviated as UHF. See: **EHF, HF, VHF, VLF**

unbalanced line - A feedline in which one conductor, the shield, is designed to be at ground potential. The most common example in Amateur Radio is coax cable.

Uncle Charlie (*slang*) - The Federal Communications Commission.

uncoordinated repeater – An Amateur Radio repeater station operating without the approval of any recognized frequency coordinating group.

uninterruptable power supply - A battery back-up power system that can provide alternating current in the event of loss of commercial power. Most modern versions also include protection for equipment from power line transient spikes as well. Abbreviated as UPS.

Universal Licensing System - The Federal Communications Commission's database and systems for processing application filings and more for all wireless services in the USA, including Amateur Radio. Abbreviated as ULS.

Universal Radio – A vendor of Amateur Radio equipment and supplies. Visit: **http://www.universal-radio.com/**

unkey, un-key (*slang*) – To release the push-to-talk switch on a microphone to cease transmitting. See: **key, key-up**

unun – A component designed to couple an unbalanced antenna of one impedance to an unbalanced feed line of a different impedance with as small a mismatch as possible. The typical ratios include 1-to-1, 4-to-1, and 9-to-1. Example: A feedline has an impedance of 200 ohms and we want to feed a 50-ohm dipole with it. A 4-to-1 unun would be chosen for the job. See: **balun, match**

UP (*CW abbrev.*) (*slang*) – "Listening up in frequency." An indication by a highly sought DX station that the operator will be listening up in frequency for calls. This keeps the large number of callers from interfering with the DX station on his transmitting frequency. Example: "I'm listening up five." See: **DN, down, DWN**

uplink - The frequency on which a user transmits to a repeater station or satellite. See: **downlink**

upper side-band, upper sideband - The frequencies on a carrier that are higher than the carrier frequency, but that contain power as a result of the modulation process. Operators using single-sideband can choose to operate either lower or upper sideband. Typically and by convention, LSB is used on 160, 80, 60 and 40 meters. USB (upper sideband) is used on all other bands. Abbreviated as USB. See: **lower side-band, LSB, USB**

UPS – Uninterruptible power supply. See: **Uninterruptible power supply**

UR (*CW abbrev.*) – "Your," "You're."

URS (*CW abbrev.*) – "Yours."

USB – 1. Upper side-band. See: **upper side-band**
2. Universal Serial Bus. Cables, connectors and communications protocols for connection, communication, and power supply between computers and other electronic devices, such as Amateur Radio transceivers.

UTC - Coordinated Universal Time. See: **Coordinated Universal Time, Greenwich Mean Time, Zulu time**

V

Victor

Di – di – di - dah

V – A common symbol for volt, a unit of electromotive force {EMF}.

vacuum tube – An electronic device consisting of a system of electrodes arranged in glass or metal envelope from which most of the air or other gas has been removed. Such devices were once commonly used in radios, televisions and other components but have been mostly supplanted by the transistor except in high-power amplification uses.

vacuum tube voltmeter - A voltmeter employing vacuum tubes. Abbreviated as VTVM. This type meter is helpful because it has a very high input impedance. This makes a VTVM valuable for measurements in circuits from which only very small currents can be drawn without altering the voltages being measured.

valve (*antiquated term*) – A vacuum tube. The term is still used in Europe and especially Great Britain.

vanity call - An Amateur Radio call sign that has been specifically requested by and issued to the operator who holds it. With some exceptions, a call sign that is currently not issued to anyone may be requested, but certain combinations are not available to all classes of license. The vanity program is administered by the Federal Communications Commission. Visit: **http://www.arrl.org/vanity-call-signs** See: **call letters**

variable frequency oscillator - A circuit within or attached to a transmitter or transceiver that varies the frequency on which the radio transmits. This lets the operator move within the bands covered by the radio and for which he or she is licensed. Abbreviated as VFO.

VE – 1. Volunteer Examiner. See: **Volunteer Examiner**

2. The call sign prefix for many Canadian Amateur Radio stations.

3. An Amateur Radio operator from Canada. Most Canadian call signs begin with the letters "VE." Example: "I had a nice chat on 40 meters with a 'VE' from Toronto."

VEC - Volunteer Examiner Coordinator. See: **Volunteer Examiner Coordinator**

velocity factor - The speed at which radio-frequency energy travels through particular conductors or feedline designs, expressed as a percent of the speed of light through a vacuum.

vertical antenna – An antenna with a radiating element that is vertical to the ground or other surroundings, and usually with radial elements arrayed beneath it in a spoke pattern.

vertical polarization – 1. An electromagnetic wave that has its electrical lines of force perpendicular to the ground.

2. A state in which an antenna has its element or elements perpendicular to the ground beneath it. See: **horizontal polarization**

very high frequency – The portion of the radio frequency spectrum between 30 and 300 megahertz. Abbreviated as VHF. See: **EHF, HF, VLF, UHF**

very low frequency - The range of radio frequencies lower than 30 kilohertz. Characterized by very long wavelengths. Long used for military communications with submerged submarines. Abbreviated as VLF. See: **EHF, HF, UHF, VHF**

VFO - Variable frequency oscillator. See: **variable frequency oscillator**

VHF - Very high frequency. 30-300 MHz-range signals. See: **very high frequency**

Visalia (*slang*) – Commonly used name for the International DX Convention held annually in Visalia, California. Alternately sponsored by Northern California DX Club and Southern California DX Club, the event is primarily for DXers and contesters. Visit: **http://www.dxconvention.com/**

VLF – Very low frequency. See: **very low frequency**

VOA – Voice of America. See: **Voice of America**

voice keyer - A device that can store and transmit pre-recorded voice messages. Especially useful in radiosport or for calling CQ. See: **CQ, radiosport**

Voice of America - A group of high-powered international shortwave stations operated by the State Department of the USA. Abbreviated as VOA.

voice operated relay - A circuit or component that senses the presence of sound through a microphone and turns on the carrier in a transmitter/transceiver to transmit. When there is a pause, the relay opens and the carrier is dropped. Usually abbreviated as VOX and pronounced "vocks." It is most often used in single sideband (SSB) operation. See: **anti-VOX**

voltage - The amount of difference in potential energy between two points in an electrical circuit. Sometimes called electromotive force or EMF. The unit of measurement of voltage is the volt. Usually abbreviated as E. See: **EMF**

voltage standing wave ratio – Abbreviated VSWR. Virtually the same as standing wave ratio (SWR). See: **standing wave ratio, SWR**

voltmeter – A device for measuring the voltage between two points in a circuit.

volt-ohm meter - A test instrument that can be used to measure current, voltage, resistance, electrical continuity, and other parameters in an electrical circuit. Abbreviated as VOM.

Volunteer Examiner - A person authorized to administer Amateur Radio license examinations. Abbreviated VE. See: **exam session**

Volunteer Examiner Coordinator - An Amateur Radio organization empowered by the Federal Communications Commission to recruit, organize, regulate and coordinate Volunteer Examiners and to assure the integrity of Ham Radio license examination sessions. Abbreviated VEC. See: **VE, VEC, Volunteer Examiner**

VOM - Volt-ohm meter. See: **volt-ohm meter**

VOX - Voice operated relay. See: **voice operated relay**

VSWR – Virtually the same as standing wave ratio (SWR), though it is specifically referring to the ratio between the highest voltage along a transmission line to the lowest voltage found at another point on the same line. Most simply refer to this value as SWR. See: **standing wave ratio, SWR**

VTVM - Vacuum tube voltmeter. See: **vacuum tube voltmeter**

VY (*CW abbrev.*) – "Very."

W

Whiskey

Di – dah – dah

W1AW – The Amateur Radio station of the American Radio Relay League in Newington, Connecticut. In addition to allowing for regular operating by staff and visitors, the station also broadcasts Official Bulletins, runs regular code practice sessions, and more. The call sign originally belonged to one of the founders of ARRL, Hiram Percy Maxim. See: **ARRL**, **Official Bulletins**

W4RT Electronics – A vendor of Amateur Radio equipment and supplies. Visit: **http://www.w4rt.com/**

WAC - Worked All Continents. An operating award for making confirmed contact with stations on each of the world's continents. Administered in the USA by the American Radio Relay League.

walkie-talkie (*slang*) - A small, hand-held, battery-powered transceiver, usually for the VHF and/or UHF frequencies. Sometimes called a handi-talkie, HT, or brick. See: **brick**, **handie-talkie**, **HT**

wallpaper (*slang*) - QSL cards, awards and special event certificates, and other items an Amateur Radio operator might proudly hang on his wall for others to see.

wall wart (*slang*) - A small switching power supply unit for low-power equipment that plugs into a standard AC wall outlet. These devices are also notorious for generating electrical noise and interference.

WAN (*slang*) – "Worked all neighbors." See: **worked all neighbors**

WARC - World Administrative Radio Conference. See: **World Administrative Radio Conference, ITU**

WARC bands (*slang*) - An expression to indicate the additional three bands that were allocated to the Amateur Radio service in most of the world at the WARC conference in 1979. Those bands are 12, 17, and 30 meters. Note that by gentlemen's agreement, no contests/radiosport events are held on these three bands. See: **World Administrative Radio Conference**

WAS - Worked All States. See: **Worked All States**

waterfall (*slang*) - A display used with digital modes that is made up of horizontal lines moving down the computer monitor screen, resembling a waterfall.

watt – The unit of measurement of power.

wavelength - The distance between successive crests of a wave, and especially between the same points in sound waves or electromagnetic waves. Typically, in the case of radio (electromagnetic) waves, the unit of measurement is meters.

WAZ – Worked All Zones. See: Worked All Zones

WBØW – A vendor of Amateur Radio equipment and supplies. Visit: **http://www.wb0w.com/**

Weak Signal Propagation Reporter – Abbreviated WSPR. Pronounced "whisper." A computer program used for digital over-the-air communication between Amateur Radio stations. The program is especially helpful with low-power transmissions over difficult propagation paths on the MF and HF bands. See: **digital modes, HF, MF**

WEFAX - Weather facsimile, satellite images and photographs transmitted by government weather satellites in orbit. Many Hams enjoy finding and downloading this data.

wet shoestring (*slang*) – Derogatory term for a very poor antenna.

WFWL (*CW abbrev.*) (*slang*) – "Work first, worry later." An expression used by DX chasers when the validity or the true identity of a station is in doubt. See: **pirate, slim**

whip antenna (*slang*) – A vertical antenna whose single element is a flexible rod or tube. Sometimes called a "stinger," and sometimes has an attachment to adjust its resonant frequency that is called a "stinger." See: **stinger**

WIA – Wireless Institute of Australia. The national Amateur Radio organization of Australia. Visit: **http://www.wia.org.au/**

wide-range antenna tuner - A matching device between transmitter and feedline that can compensate for very large impedance mismatches. See: **matchbox, tuner**

wilco (*slang*) (*antiquated term*) – Literally "I will comply." Once commonly used as, "Roger wilco," meaning "I understand and will comply."

windom antenna – An off-center-fed dipole antenna designed to present a reasonable match on several Amateur Radio bands. See: **OCF, off-center fed antenna**

window (*slang*) - A range of frequencies set aside by gentlemen's agreement to allow for foreign Amateur Radio stations to call CQ and work other stations around the world while those in the United States and Canada refrain from other types of activity there. The DX stations may also invite U.S. and Canadian stations to call them. Sometimes called a "DX window." Example: 3.790 – 3.800 megahertz is the 75 meter DX window. Visit: **http://www.bandplans.com/**

window line -   A type of antenna feedline using two parallel conductors encased in plastic insulation. Holes have been punched in the middle area between the two conductors to reduce weight and so the line does not fluctuate so much in the wind. See: **balanced line, ladder line, open wire line**

Winlog32 – A popular software program for logging Amateur Radio on-air contacts and for use in radiosport events. The program is available for free download. Visit: **http://www.winlog32.co.uk/**

wireless (*antiquated term*) – A term once used to identify radio communications as opposed to wired means such as by telegraph.

Wireless Institute of Australia – The National Amateur Radio organization of Australia. Abbreviated as WIA. Visit: **http://www.wia.org.au/**

work (*slang*) - To carry on a valid two-way radio contact with another Amateur Radio station. Example: "I see in the log that we worked back in 2007."

Worked All Continents - An operating award for making confirmed contact with stations on each of the world's continents. Administered in the USA by the American Radio Relay League. Visit: **http://www.arrl.org/wac**

worked all neighbors (*slang*) – A fictitious award for a Ham Radio operator with a serious TVI or RFI problem, meaning he has interfered with all the people living nearby. Abbreviated as WAN. See: **RFI, TVI**

Worked All States - An operating award for making and confirming contact with Ham Radio operators in each of the fifty United States. Administered by the American Radio Relay League. Abbreviated as WAS. Visit: **http://www.arrl.org/was**

Worked All Zones - An operating award for making and confirming contacts with other licensed Amateur Radio operators from each of 40 zones, geographical areas designated by *CQ Magazine*, the entity that administers the award. Visit: **http://www.cq-amateur-radio.com/cq_awards/cq_waz_awards/index_cq_waz_award.html** See: *CQ Magazine*, **CQ zones**

World Administrative Radio Conference - A technical conference of the International Telecommunication Union (ITU) where delegates from member nations of the ITU meet to revise or amend the international radio communications treaties and agreements. A major part of the conferences is determining assignment of parts of the radio spectrum to the various parties who use them. Abbreviated as WARC. See: **ITU**

World Radio Laboratories – A former manufacturer of Amateur Radio equipment, best known for its line of Globe transmitters. Abbreviated as WRL.

Worldwide Radio Operators Foundation - An independent organization devoted to the skill and art of radio operating. The group focuses on operating with emphasis on radiosport and how such activities can improve communications ability and station competency. Abbreviated as WWROF. Visit: **http://wwrof.org/**

worm burner (*slang*) - An antenna system so near to the ground that it tends to have most of its energy absorbed by the earth beneath it. See: **cloud warmer**

wouff hong (*slang*) – A mysterious and highly dreaded weapon that was supposed to be used on any Ham who insisted on exhibiting poor operating procedures on the air. Created by Hiram Percy Maxim, the original holder of the W1AW call sign and a co-founder of the American Radio Relay League. Visit: **http://www.arrl.org/ham-radio-history**

WPM (*CW abbrev.*) - Words per minute. How fast Morse code is being sent or received.

WPX – World prefix contest. A radiosport event in which operators attempt to work as many stations with unique call sign prefixes as possible within a set time period, typically a weekend each year for phone modes and a weekend for CW. Sponsored by *CQ Magazine*. Visit: **http://www.cqwpx.com/** See: **prefix, radiosport, suffix**

WRK (*CW abbrev.*) – "Work."

WRKG (*CW abbrev.*) – "Working."

WSJT - A computer program used for weak-signal radio communication between Amateur Radio operators. See: **digital, digital modes, JT-65**

WSPR - Weak signal propagation reporter. Pronounced "whisper." A computer program used for digital communication between Amateur Radio stations. The program is especially helpful in sending and receiving low-power transmissions over difficult propagation paths on the MF and HF bands. See: **digital modes, HF, MF**

WWROF – Worldwide Radio Operators Foundation. See: **Worldwide Radio Operators Foundation**

WWV – A radio station operated by the National Institute of Standards and Technology, an agency of the U.S. Department of Commerce. The station offers a frequency reference source by allowing anyone to calibrate to their transmit frequency. They also send highly accurate time signals and radio propagation reports on 2.5, 5, 10, 15 and 20 megahertz, and at times on 25 megahertz, all from the location in Ft. Collins, Colorado. Visit: **http://www.nist.gov/pml/div688/grp40/wwv.cfm** See: **WWVB, WWVH**

WWVB – A radio station operated by the National Institute of Standards and Technology in Ft. Collins, Colorado. The station broadcasts time signals that are accessed by radio clocks all over the world, which synchronize to WWVB's highly accurate telemetry. Visit: **http://www.nist.gov/pml/div688/grp40/wwvb.cfm**

WWVH – A sister station to WWV, the U.S. National Institute of Standards and Technology's shortwave radio time signal station. This station broadcasts from the island of Kauai in the state of Hawaii. Visit: **http://tf.nist.gov/stations/wwvh.htm** See: **WWV**

WX (*CW abbrev.*) – "Weather."

WX4NHC – The Amateur Radio station at the National Hurricane Center in Miami, Florida. Visit: **http://w4ehw.fiu.edu/**

X

X-ray

Dah – di – di – dah

X – Symbol for the electrical property known as reactance.

XCVR (*CW abbrev.*) – "Transceiver."

XIT - Transmit incremental tuning. A control that allows the operator of a transceiver to change the transmit frequency while leaving the receiver on its original frequency. See: **RIT, receiver incremental tuning**

XLR connector – A type of audio connector typically found on professional audio equipment.

XMIT (*CW abbrev.*) – "Transmit."

XMTR (*CW abbrev.*) – "Transmitter."

XTAL (*CW abbrev.*) – "Crystal." See: **crystal**

XVTR – Transverter. A circuit in or attached to a transceiver or transmitter that allows the radio to operate on other bands. This capability is typically employed so a transceiver designed only for HF can be used on VHF or UHF bands.

XYL (*CW abbrev.*) (*slang*) – "Wife." "Ex-young lady."

Y

Yankee

Dah – di – dah – dah

Yaesu – A major manufacturer of Amateur Radio equipment, headquartered in Japan. Visit: **https://www.yaesu.com/**

Yagi - A beam-type directional antenna consisting of a dipole and two or more additional elements, including a slightly longer reflector and a slightly shorter director. Electromagnetic coupling among the elements maximizes signal gain on both transmit and receive in the direction of the director.

YL (*CW abbrev.*) (*slang*) – 1. "Young lady," an unmarried female Ham. 2. Any unmarried female.

YL Radio League - An organization for licensed female Amateur Radio enthusiasts. Abbreviated as YLRL. Visit: **http://www.ylrl.org/** See: **YL, XYL**

YLRL – YL Radio League. See: **YL Radio League**

YR (*CW abbrev.*) – "Year."

Z

Zulu ("ZOO – lew")

Dah – dah – di – dit

Z – The letter used to represent electrical impedance.

zed - a phonetic pronunciation for the letter "Z" used to avoid operators mishearing the letter as "C" or other letter. It is commonly used on phone modes in the place of "Z," especially in Australia, Canada, New Zealand, and Great Britain. Example: An operator with the call sign W4ZHR might give his call as, "W 4 zed H R."

zero beat - Adjusting the frequency of a transmitter so it is on precisely the same frequency as another station.

zepp antenna – An antenna consisting of a single wire, one-half wavelength long on its design frequency, and fed from one end. This antenna was developed to be used as an antenna for zeppelins and other lighter-than-air craft from which it could be reeled out when in use and then reeled back in for landing or docking.

Z-signals – Three-character codes, similar to the Amateur Radio Q-code but beginning with the letter "Z." Used on CW, primarily in the military and in the Military Affiliate Radio System. Visit: **http://www.radiotelegraphy.net/zsignals.htm** See: **MARS, Military Affiliate Radio System, Q-code**

zulu (time) – Military term for Coordinated Universal Time. Means the same as Greenwich Mean Time and Coordinated Universal Time, which is the time at 0-degrees longitude, which passes through Greenwich, England. Represented by the letter "Z" after the time in 24-hour format. Example: 3:40 PM UTC is 1540Z. See: **Coordinated Universal Time, UTC, Greenwich Mean Time**

NUMBERS and PUNCTUATION

Ø – The number zero with a slash through it in order to distinguish a zero from the capital letter "O." (To create this character on most keyboards, hold down the Alt key while pressing Ø 2 1 6 on the numeric keypad.)

5/8 wave (*slang*) – An antenna that is 5/8 of a wavelength long. Usually applies to a VHF or UFH vertical radiator with a loading coil at its base.

1-by-1 call sign – A special temporary call sign consisting of a single letter, a number, and another single letter, issued to special event stations or for other distinct purposes. Example: The call sign N9N was issued for the special event station commemorating the fiftieth anniversary of the journey to the North Pole by the submarine USS *Nautilus*. Visit: **http://www.1x1callsigns.org/**

10 code - A series of abbreviations created originally for public safety communications but that were adopted and modified considerably by Citizens Band operators. The 10 code is not in use by Amateur Radio and is frowned upon when used by newcomers.

10 – 10 – Ten Ten International. See: **Ten Ten International**

11 meters (*slang*) – Another name for the Citizens Band. 11 meters is the wavelength of the 27 megahertz range assigned to that service. This was once a Ham band but was taken away to create the new CB service. Some resentment lingers.

33 (*CW abbrev.*) (*slang*) – "Love sealed with friendship and mutual respect between one young lady (YL) Amateur Radio operator and another." Adapted officially by the YLRL. See: YL, **YL Radio League**, XYL, **YLRL**

5-2 (*slang*) – The national 2-meter simplex calling frequency of 146.52. Example: "I'll be listening for you on 5-2." See: **calling frequency, simplex**

73 (*CW abbrev.*) (*slang*) – "Best regards." Though intended for CW use, "73" is used extensively on voice and digital modes.

73 *Magazine* – A former Amateur Radio magazine, now defunct.

88 (*CW abbrev.*) (*slang*) – "Love and kisses."

92 code - An early numerical code adopted for radio telegraphers. "88" and "73" were originally from that code but none of the other numbers remain in use in Amateur Radio. See: **73, 88**

807 (*slang*) - A beer. Named for a popular transmitting tube of the past that resembled an upside-down beer bottle.

9913 - - Common type of coaxial cable used by Hams as feedlines for antenna systems. See: **LMR-200, LMR-400, RG-6, RG-8, RG-8X, RG-58, RG-59**

? (*CW abbrev.*) – "Repeat."

/ (*CW abbrev.*) - "Slash" or "front slash." In Amateur Radio, typically sent before a designation that the station is operating portable or maritime mobile. Sent as dah-di-di-dah-dit.

ABOUT THE AUTHOR

Don Keith is an award-winning broadcaster, a best-selling author, and has been a licensed amateur radio operator since 1961. He was first licensed as WN4BDW in 1961, earned his General class license later that year, and then became an Amateur Extra class licensee in the mid-1970s, changing the call sign to N4KC.

He was twice named *Billboard Magazine*'s "Broadcast Personality of the Year," won every major broadcast journalism award in his state from the Associated Press and United Press International, and was an on-air personality, journalist, station owner, program director, and manager in a broadcasting career that spanned over two decades.

Don published his first novel, *The Forever Season*, in 1995. It has remained in print continuously since and was named "Fiction of the Year" by the Alabama Library Association. His more than two dozen other published works, fiction and non-fiction, cover such topics as NASCAR racing, broadcasting, college sports, submarines, biography, and World War II history.

Don lives in Indian Springs, Alabama, with his wife, Charlene, has three grown children, and three grandchildren. He operates all the shortwave amateur radio bands as well as VHF and uses most modes, including CW, SSB, PSK31 and FM. He enjoys DXing, contesting, antenna experimenting, and just plain rag-chewing.

Don's author web site is **www.donkeith.com**. His Amateur Radio web site is **www.n4kc.com** and features a number of stories and articles about the hobby and of interest to Hams.

Enjoy this book? You would certainly like Don's other Ham Radio book, "Riding the Shortwaves: Exploring the Magic of Amateur Radio." See more and read a sample at **http://www.donkeith.com/hamradio/riding-the-shortwaves**